INDEPENDENT CONTRACTOR

--

WHY AND HOW.

Sheldon "Shelly" Waxman, J.D.

ISBN: 978-1461176985

INDEPENDENT CONTRACTOR

—

WHY AND HOW

CONTENTS

APPENDIX

ACKNOWLEDGEMENT

This book is dedicated to my Mom (who overcame some terrible hardships to become a successful independent contractor— RIP), my Dad—a professional violinist (gone, but never forgotten), Josiah (my lovable bright son), Zoe (my adorable smart daughter), all my other good friends, and to all trying to free themselves from the wage-slave employment payroll system.

"The natural liberty of man is to be free from any superior power on earth, and not to be under the will or legislative authority of man, but to have only the law of nature for his rule."
John Locke

"If a man empties his purse into his head no one can take it away from him. An investment in knowledge always pays the best interest."
Benjamin Franklin

PREFACE

The following is what some very savvy people have said about independent contracting:

SAM WALTON :
"We can get beyond old adversarial relationships and establish win-win partnerships with our suppliers and our workers, which will leave us with more energy and talent to focus on the important thing, meeting the needs of our customers. All this requires overcoming one of the most powerful forces in human nature: the resistance to change. To succeed in this world, you have to change all the time." *Made in America*

PETER DRUCKER:
"The possessor of knowledge owns his own 'tools of production' and has the freedom to move to wherever opportunities for effectiveness for accomplishment and for advancement seems greatest. The knowledge society will inevitably become far more competitive than any society we have yet known—for the simple reason that with knowledge being universally accessible, there are no excuses for non-performance. There will be no 'poor' countries—there will only be ignorant countries. And the same will be true for individual companies, individual industries and individual organizations. It will be true of individuals too." *Managing in a Time of Great Change*

GERRY SPENCE:
"The new and most powerful union of all will be a union of one: one man, one woman, one worker with special skills, an inquiring mind, an independent attitude. In the new-age workplace the worker will no longer be a slave. He will enter the place of work voluntarily to do a job for a price, his price. He will leave as he chooses. He will cherish his freedom, which is his security. He cannot be lured into the trap. The master cannot own him." *Seven Simple Steps to Personal Freedom*

WHY AND HOW

If you have been told that it is too dangerous to be an Independent Contractor or to hire them, DON'T BELIEVE them. It is now easier than ever to be or use Independent Contractors. The IRS is mostly on our side. They haven't given up though with their recent unsuccessful attempt to have 1099's issued for all payments to everybody. The CPA's and accountants who tell you not to use or be independent contractors say that out of ignorance of the rules and because of fear of the IRS and, of course, because they will lose the business income they make from doing payrolls.

I also provide a CERTIFICATION procedure whereby you or your workers are guaranteed to be properly operating as Independent Contractors for IRS and other purposes.

E-Mail me at   sheldonw72@gmail.com

See also   www.independentcontractor.info

This Certification procedure is a FOOL PROOF method of stopping the IRS and other governmental agencies from even attempting a challenge.

PART I

INTRODUCTION

CHAPTER 1: DEFINITIONS

It is hoped that this Book will provide you with the information necessary to attain the Independent Contractor treatment that you are seeking. To assist in understanding what is involved, it is necessary to provide you with the following abbreviations and definitions:

IC=Independent Contractor—a self-employed, independent business person;

SP=Service Provider—The one who provides a service to the Service Recipient;

SR=Service Recipient—the person/company who contracts with you (the SP) to do work and provide your services;

ES=Employment status/system—the employment system that depends on the payment of withholding and other taxes and other onerous state and federal regulations, the system you are trying to transform away from;

CLFT=The 20 common law factor test—which is used by the IRS to determine if you are an IC;

SHT=The Safe Harbor Test—a test, which if you are covered by it will immunize you from any reclassification procedures. The SHT is only available to the companies that hire SP's.

CW=Contingent Workers. This does not include IC's. These are varying forms of leased or agency workers, who are the outsourced personnel that have become popular with companies. It is not recommended that you become one of these or use these, although some use it as a step in the progression to become an IC. It is not recommended because the SP just becomes a wage slave to the hiring agency and the SR pays a lot to have the agency handles all of the ES regulations, payroll and taxes.

CHAPTER 2: THE MASTER-SLAVE RELATIONSHIP

We've turned into a nation of whiners, where employees spend their days hoarding imagined insults and injuries for future lawsuits, and employers are forced to offer continued employment to unsuitable employees. The old adage of an honest day's work for an honest day's pay is no longer valid.

The employer-employee relationship has become an adversarial union based on mutual distrust and suspicion. There used to be an unspoken agreement that was made when you were hired and exchanged a handshake with your new boss. You went to work every day—barring some specific instances—and in return, you received a living wage. I don't know when the time-honored bond between the employed and the employer began to decay, but it seems to be getting worse. Independent Contracting is the way out of this mess. It's time has come.

Today's employees expect to start at the top of the business ladder, calling the shots and making the rules, despite their lack of experience, knowledge or skills. Prospective employees now come to the employment table with a l ist of wants. They don't ask. They demand to receive benefits and considerations that were formerly reserved for trusted longtime employees. Many bring no pertinent skills to the position they are applying for. They want a job but not necessarily work. Employees have no incentive to develop a sense of company loyalty.

If an employee isn't working out, it almost takes an act of Congress to relieve them of their duties lest the employer be accused of "discrimination."

A paper trail of documentation leading back to the dawn of time must now be kept, outlining the employee's transgressions. Even the old "three strikes and you're out" practice no longer applies. Instead of ridding themselves of ill-suited employees, employers must now offer a plethora of professional services to help rehabilitate the erring employee, etc..

We need a work ethic that allows for mutual respect between workers and those for whom they work.

CHAPTER 3 A BUSINESS OF ONE

It started around the time the Civil Rights Act of 1964, created the Equal Employment Opportunity Commission that was passed as part of Lyndon Johnson's Great Society program. Of course the Federal and State Governments had been meddling in the contract between employee and employer long before this with withholding tax requirements, social security, unemployment benefits, workman's compensation, union check-off and fair labor standards. But with the creation of the EEOC, personnel practices of private companies became a favorite target of socialistically minded legislators.

It has reached such a stage that the concept of private employment is now a joke. The Federal and State Governments now control employment. "Personnel Departments" are now called (Human Resource Departments)—further signifying the encroachment of socialistic concepts into the work relationship. Of course, each State has gotten into the act with their "packages" of restrictive laws.

The following are just some of the federal restrictions that have come about since 1964:

a. Pensions-ERISA—Employee Retirement Income Security Act.

b. INS (Immigration and Naturalization Service) regulations.

c. ADEA—Age Discrimination in Employment Act.

d. OSHA—Occupation Safety and Health Act.

Moreover, private causes of action have sprung up, having been allowed by the Courts. to curtail the right of employers to hire and fire whomever they wanted. This litigation game has become a big industry for the legal profession and has spun off a specialty known as "employment law." It is a specialty that didn't exist before all this legislation.

Will this trend continue? I don't think it can. It has become too costly and too inhuman. It has to implode, eventually. The excessive regulation will topple as business owners are unable to expense the required paper work. The only thing better than the collapse of the regulations, of course, would be elimination of the Internal Revenue Code, which isn't about to happen anytime soon. In the meantime, Independent

Contracting is the only way to insure a limitation on how much you pay to the governments.

People who do work for other people have come to expect employment as a right and not a contract. They expect the benefits that the laws have mandated as their due, whether the company is profitable or not. This cannot continue and it won't because there is a relatively new concept on the horizon. It is the "business of one" where each of us is his own company and each of us is his own business person. It will become the new relationship once it is understood how it can be accomplished without getting into trouble with the government.

The history of employment can be traced over its development of thousands of years. It runs from Master/Slave to Master/Servant (the indentured slave) to Employer/Employee. Employer/Employee is a master servant concept where the employee is an indentured slave (controlled by the employer as a matter of definition) but only for 8 (or however many) hours a day that the employee is under the employer's thumb.

The new kid on the block—but not really because the concept dates back to the Guilds of England where the tradesmen were uncontrolled in the manner of their work—is the independent contractor. Traditionally, the professions (lawyers, doctors, etc.) have also acted as independent contractors. The concept of a business of one is the self-employed relationship or the SR/SP relationship.

By its very definition, it is not a relationship controlled by the hiring agent, who may only have control over the end product produced by the independent contractor. Independent contractors and those who hire them are not subject to the laws imposed on the employment relationship.

The law of contract is the only law that applies in an IC relationship because it is contractual and not legally imposed, as is the case with employment. In the late 1960's and early 1970's, the Federal Government attempted to destroy the concept of independent contractors by forcing those who hired them to withhold the taxes required of employers—the dreaded "Taxation at its Source". Taxation at its Source (withholding) makes employers collection agents of the government.

Withholding—the idea of a fellow with the unlikely name of Beardsley Rummel—was supposed to be only an emergency short-term World War II measure. Milton Friedman, economist and Nobel Laureate, who was partially responsible for its enactment, has stated that it was the worst mistake of his lengthy career and he was ashamed of his involvement in its enactment. Of course the end of the War did not end the withholding system.

In the 1970's the IRS started a special project aimed at turning everybody into an employee for withholding tax purposes. Massive assessments were levied against companies who were only left off the hook, if they agreed to contribute to the Nixon re-election committee and treat their workers in the future as employees. This is an unknown result of the Watergate scandals.

The Federal Government's attempt to destroy independent contractors failed. It is a long story, but suffice it to state that the government is no longer interested in the fight. The matter has evolved so that the governments (the politicians, not the bureaucrats) recognize that independent contractors are necessary to a growing economy. It is just what we need in the Stagflation economy that is soon to come.

Everybody is outsourcing to independent contractors. If the practice were ended some of the world's largest companies would go broke. Not all workers will qualify for independent contractor status but it is a positive work arrangement in most cases. For the Service Recipient the independent contractor set up means that it will build alliances and build their businesses not their overhead.

The following advantages for the SR are sure to be part of the transformation:

1. Flexibility
2. Specific expertise
3. Cost cutting-monetary advantages
4. Tax benefits
5. Inapplicable regulations
6. Unfounded lawsuits and adversarial relationships
7. Forced continued employment of unsuitable workers
8. Elimination of expensive benefits
9. Elimination of costly accounting and mandated record keeping
10. Discontinue demeaning personnel practices, such as drug testing
11. Ability to rid oneself of workers who lack necessary skills
12. Elimination of negligence liability under most circumstances

For the SP, the following are the advantages:

1. Flexibility/control of benefits
2. Greater pay
3. Job security
4. Be own boss
5. More concern for family obligations
6. Tax benefits—Schedule C
7. Own decisions regarding insurance and pensions

The government recognizes that its welfare programs, as we know them, will end, if everybody becomes wealthy and being your own business is a way to become wealthy. Furthermore, the reason often cited for the battle to eliminate independent contractors no longer exists. Independent contractors pay their taxes and the IRS now cross-matches 1099 forms.

Moreover, the government is having problems with the payroll system in that many companies are starting up, and collecting the taxes from the employees but not paying their withholding taxes. They take the money from their employees, fold up, disappear and never pay over the money.

It has been estimated in the Senate Hearing on the subject that over $50 billion has been lost to the government this way and the figure is rising. Since the employer has taken the money from the employees, the government still has to credit the employee with the withheld money. Therefore, it is a double loss to the government because they have to payout the social security and unemployment money they never receive.

But why is this subject still mostly a secret? Well, there is an inbuilt reaction to change. There are those who will counsel not to change because their self-interest is at stake. It has been estimated that a changeover of a business to an independent contractor operation will save 20% on accounting overhead. The only paperwork required of the Service Recipient is the 1099 form at the end of the year (and this is not required for payments to corporations) providing the total amount paid is more than $600.

Obviously the accountants, who make their money filling out the endless forms required for the payroll system, will not benefit from any change. Banks will no longer receive their automatic payroll deposits. Moreover, the lawyers, who are naturally resistant to change, will tell their clients that to change will reap them the government's winds of wrath.

The truth is that new rules have resulted in the federal government not involving itself in payroll enforcement activity related to the IC issue since 1996. The problem will not be with the Feds, if it is properly done. And it must be emphasized that there are rules that have to be strictly followed. Problems may exist with State regulations, which vary from state to state. However, if one is an independent contractor for federal law purposes, it is more than likely to be so for state law purposes.

PART II

FOR INDIVIDUALS

C HAPTER 4: WHY BE AN IC

Being an IC is not for everybody. If you have the knowledge and desire to be your own boss—it is the only way to go. Why? Because the advantages of IC and self-employment status far outweigh those of the ES, not only in compensatory advantage but, also, the non-material freedom advantages that the IC status provides.

It is assumed that by purchasing this book, you are already interested in at least exploring this subject to determine whether it is for you and how to do it. Of course, we are dealing with a transformation and in all such paradigm shifts, change has come about slowly because resistance to it has been powerful, as is always the case. Those enamored with the *status quo* (particularly lawyers and other so-called tax professionals) resist with all of their might, rather than get on board to advance the clear IC trend in employment.

Along with the advantages of IC status, comes the task of being responsible for oneself. This is something that we, as proud Americans, have gotten away from. We have been taught that it is the responsibility of the government or our "employer" to care for us. A surge of self-reliance has come about recently because we all know there is too much Big Brother.

This is the concept of a "Business of One". Gerry Spence called it a "Union of One" but I like Business of One better. This is what the IC business is all about—you, as an IC, must take care of yourself. That means you must insure that you are adequately paid so that you can pay for what you need that the SR (unless provided for in the IC agreement) and the government will no longer provide, such as: General liability ("umbrella") or Errors and Omissions insurance; Severance pay or building a stash; Disability insurance; Medical and Dental; Pensions; and Continuing education.

The advantages and benefits to which you are entitled as an IC, both pecuniary and in lifestyle, are as follows:

1. You will be able to take the tax deductions from gross income not allowed to employees. See your local tax advisor.

2. You will only have to make quarterly tax payments but even those under certain circumstance do not have to be made. Ask your tax advisor. Even so, the penalty for not doing so is small.

3. You will build up your assets and provide yourself with the necessary specialized training to assist you in your business, the purchases of which will be tax deductible, increasing your productivity and value to your customers and establishing that you are entitled to increasing amounts of pay.

4. You will be able to add to your customer base, thereby ensuring continued work in the event of slowdowns and acquire greater job security.

5. You will be able to arrange your own insurance needs, according to your own preferences.

6. You will have the flexibility to determine when you want to take your vacations and time off work.

7. You will be your own boss.

8. You will determine your own retirement plans and will be eligible for a Keough Plan (HR10) which will enable you to deduct 25% of your gross income up to $30,000 per year to go into your retirement account. Additionally you will be in full control as to how your retirement funds are invested. You will also be eligible for various IRA plans. For further information on this subject contact your local financial advisor.

The biggest benefit that an IC enjoys is in most cases, although not a necessity as far as the rules go, is the ability to work for more than one SR, thus ensuring continuing pay in the event one SR reduces staff, the task is finished, or the SR goes out of business

The trend is away from the ES, especially during periods like we are undergoing. Along with the changing landscape of the enormous effects of the technology revolution (the information age-considered to be our third economic revolution after "farming and "industrial") has come the changing concepts of people towards their work. You will be in control of your work rather than "the company." It is in your interest to work as hard as you can.

The ES is outdated, encumbered as it has become, with all of the laws that have been piled on it. It has now been declared criminal to post an ad that states "unemployed not accepted." That's a violation of Freedom of Speech, isn't it? But it shows how far the government has come. Human Resource Departments (formerly called "Personnel Departments"), which I sometimes humorously call "Human Relations" because they have taken over so much of corporate life. HR Departments are the ones in power now and they are dictating rules for their employees, rather than trying to save their companies money.

The ES has led to low productivity and endless work conflicts. It was meant to apply to a centralized workforce system where everybody went to work at a certain location and started and ended at a certain time with detailed instructions given as to how the work was to be done. The ES is not conducive to the decentralized work system that we are now entering.

Work has become task oriented with teams of IC specialists accomplishing a task put before them by the SR. How you, as an IC, do your work is not important, as long as the task is satisfactorily completed within parameters established for the end product within a time-frame set by the SR.

The laws engrafted onto the ES, mostly of recent origin, have guaranteed the death of that system. However, there will be a continuing need for employees, such as secretaries, unless they have their own secretarial business and work for more than just one company in which case they would be IC/SP's on their own.

Companies need certain people to be controlled as to hours and performance of jobs. The lack of the right to control an IC, as hereafter explained, is the difference in the work of an IC, as against an employee. Certain people have a need to work in such a controlled environment where they are told how, when, and where to perform their work. We assume you are not one who wants to be an employee or else you would not be interested in this Book.

The laws meant to protect the employee from all manner of conceivable alleged injuries in fact produce a slave and conflict oriented mentality toward work. These laws do not apply to the IC status.

INDEPENDENT CONTRACTOR

I hope you find this Book an easy way to understand what has been made to appear to be a complicated subject. Simplicity is what we are striving for because so many others have attempted to make it the subject of fear, as a barrier to making the transformation. Some of these alleged "how to do it" books contain hundreds of pages of complicated explanations of what is really a simple matter.

CHAPTER 5: REQUIREMENTS FOR IC STATUS

If you work for somebody who has the right to control your work, you are an employee. In the IC situation, the SR only has the right to receive a quality end product or service. The IRS has simplified its concerns and now concentrates on only 3 points of view: Behavioral, Financial, and the Relationship of the Parties in making determinations of employment status.

It is always difficult to explain a complicated subject simply. But that is the purpose of this Book. There are two laws that provide the mandatory requirements for IC treatment. One is called the 20CLFT and the other the SHT. The 20CLFT is the most complicated test known to the law because no one or combination of factors controls. The determination is made by looking at the IC's relationship to all of the factors as a whole. The 20CLFT is used by the IRS to determine whether you are an IC. State laws pretty much conform to this test in one variety or another.

In addition to the 20CLFT, Congress because it was angry at the IRS' attempt to destroy IC's in the late 1970's and early 80's, passed an interim measure then known as Section 530 of the Revenue Act of 1978, the SHT. This law was reenacted each year until 1982 when in revised form it was made permanent by the Tax Equity and Fiscal Responsibility Act of 1982.

If you are a certain type of "techie" specialist who is brokered to other companies, you were made a "statutory employee," as are others: government workers; food and laundry delivery drivers (unless they own their own business); certain types of at home workers who are supplied with the materials they use; full-time salespersons who sell goods for resale; and corporate officers. Heaven only knows why these groups of people were selected to be employees by law, and I am sure there are ways to get around these rules.

The statutory employee designation for corporate officers can be partially avoided. An IRS Chief Counsel's opinion allows corporate officers to be IC's to their corporation as will be covered in the "Dual Hat" Chapter.

Certain other occupations are "statutory IC's", such as: real estate agents, if they have a written IC Agreement with their Brokers; and direct sellers of consumer products, if they have a written IC Agreement with their suppliers. The truth is that these exempted classes were made IC's because of the power of their lobbyists and inside knowledge of the IRS' unlawful plan to make us all employees. I co-operated with them and fed them documents I had received in the Dema/Tabcor cases. See, Author's Biography in the Appendix

According to the legislative history of the 1982 Act, it is to be "liberally construed" in favor of the one claiming to be an IC. Additionally, pursuant to the Small Business Jobs Protection Act of 1996, for periods after 1996, the IRS is required to bear the burden of proof that an individual is not an IC.

The SHT is called that because if you come within its protective (in other words safe harbor) provisions, you are automatically an IC. That is the present and simply put status of the law. Now for the nitty-gritty—the absolute bottom-line requirements.

The SHT does not replace the 20CLFT; it only supplanted it—and became an umbrella of protection for IC's. It is only important for you to remember that if you qualify under the 20CLFT, that is a "reasonable basis" under the SHT and you automatically qualify. The IRS will not be able to reclassify you. The SHT applies only to companies, but they are the ones who need the protection of the SHT because the IRS rarely goes after the individual.

So the only question that remains is what are the absolutely necessary factors out of the 20 with which you will need to comply. In the 1996 IRS Training Manual, the IRS loosened its interpretation of the 20CLFT for fear of further Congressional restrictions. The IRS recognized for the first time that being an IC "can be a valid and appropriate business choice."

Certain of the 20 factors are no longer important to the IRS. Those used to be the difficult ones. These are the critical factors:

1. You offer your services to the public—advertising in a local newspaper, grocery bulletin board, flyers, yellow pages, etc. Just because

you advertise does not mean you have to accept jobs from those who may call you. Besides you should want as many customers as you can handle.

2. You must have a written agreement with each SR indicating your intent to be an IC and that you are permitted to work for other SR's, if you so desire. The particulars of this agreement are supposed to be negotiated fairly between you and the SR and can include other benefits besides pay.

3. You must own or rent your own equipment.

4. You must invoice your billings . Your pay can be hourly, daily, weekly, monthly, yearly, flat fee, bonus, incentives, etc. Flat fee payable in increments is preferable but not a necessity. A simple invoice is all that is necessary but you must invoice your SR, preferably on a periodic basis but not necessarily on the same days.

5. You should have unreimbursed business expenses reportable as deductions from Gross Income on IRS Schedule C.

6. Although not a requirement for you, the SP, but is required of the SR, the SHT requires that the SR, who retains your services, must file 1099 forms with the IRS at the end of the calendar year. Penalties for failure to file 1099 forms are negligible. The SHT also provides that you cannot be an employee and an IC for the same "period", meaning the quarterly, not the annual period.

7. Although not mandatory, it would be best for you to incorporate and sign a management services agreement with your corporation. Payments from one corporation to another do not require the filing of 1099 forms. However, in any event you must have a business presence whether it is sole proprietorship accomplished by registration with your county clerk, under the assumed name law or a partnership or any other entity, such as a limited liability company. Well that is it. It is easy.

However, if you desire additional protection, I can provide it for a reasonable fee. This would take advantage of Court interpretations that reasonable reliance on a professional's opinion constitutes a "reasonable basis," according to the SHT.

The Courts have stated:

"(The SR's) reliance on the advice of professional tax advisors (that its SP's were IC's under the 20CLFT) is sufficient to demonstrate a reasonable basis under (the SHT) for not treating its (IC's) as employees... reliance upon the professional advice rendered by the (tax

21

professional) constitutes a r easonable basis…Generally, the courts have found that reasonable cause exists where the taxpayer relied on the advice of a trusted attorney or accountant. (Citing authority)

Indeed in this regard, the Supreme Court has stated:

'When an accountant or attorney advises a taxpayer on a matter of tax law, such as whether a liability exists, it is reasonable for the taxpayer to rely on that advice. Most taxpayers are not competent to discern error in the substantive advice of an accountant or attorney. To require the taxpayer to challenge the attorney, to seek a "second opinion," or try to monitor counsel on the provisions of the Code himself would nullify the very purpose of seeking advice of a presumed expert in the first place. 'Ordinary business care and prudence' do not demand such actions." (Citing Supreme Court case)

PART III
FOR COMPANIES

CHAPTER 6: NEED TO USE IC'S

The following advantages for the SR are sure to continue to impel the IC Evolution:

1. Flexibility
2. Specific expertise
3. Cost cutting-monetary advantages
4. Tax benefits
5. Many laws and regulations do not apply to IC's.

a. Pensions-ERISA—Employee Retirement Income Security Act. *Community for Creative Non-Violence v. Reed,* 490 U.S. 730 (1989).

b. ADA-Americans with Disability Act.

c. NLRA—National Labor Relations Act—*NLRB v. United Insurance Co.,* 390 U.S. 254 (1968).

d. FLSA—Fair Labor Standards Act—*Tony and Susan Alamo Foundation v. Secy. Of Labor,* 471 U.S. 290 (1985).

e. INS (Immigration and Naturalization Service) regulations.

f. EEOC—Equal Employment Opportunity Commission— Title VII of the Civil Rights Act of 1964. *Kirby v. Swimfashions,* 904 F.2d 36 (6th Cir. 1990).

g. State Workers Compensation Acts.

h. ADEA—Age Discrimination in Employment Act. *Lorilard v. Pons,* 434 U.S. 575 (1978).

i. OSHA—Occupation Safety and Health Act—*Cochran v.International Harvester Co.,* 408 F. Supp.598 (W.D., Ky, 1975).

j. Family Leave Act

k. Mandated vacation days

6. An end to frivolous lawsuits and adversarial employee relationships

7. An end to forced continued employment of unsuitable workers

8. Elimination of expensive benefits

9. Elimination of costly accounting services and governmentally mandated record keeping

10. Discontinuation of demeaning personnel practices such as drugtesting

11. No need to keep workers who lack necessary skills

12. Elimination of negligence liability under most circumstances

13. IC's do not belong to unions—If that doesn't convince company honchos to get on board, I don't know what will do it. I think companies who don't take the leap could be criticized by their shareholders for not doing so.

CHAPTER 7: RESISTANCE TO CHANGE

As stated by Sam Walton, the founder of Wal-Mart in his autobiography, "Made in America", who made constant change the number one priority of his business: "I've made it my own personal mission to ensure that constant change is a vital part of the Wal-Mart culture itself. I've forced change—sometimes for change's sake alone—at every turn in our company's development. In fact, I think one of the greatest strengths of Wal-Mart's ingrained culture is its ability to drop everything and turn on a dime."

Accountants want to continue with payroll; they are not as concerned as are you in devising ways to save you money. They are concerned with making money themselves and payroll is very lucrative for them. Lawyers are known as the most regressive of the professions in terms of accommodating to change for a number of reasons, most of which relates to the concept that, "We have always done it this way, so why change."

These advisors will think of a million reasons why you should not change. Don't listen to them, unless they give you good reasons. IC status is legal and easy to accomplish and risk free.

Lawyers and accountants often cite the example of Microsoft's famous case as a reason not to convert. However, a reading of the case produces a different result. Realizing the cost effectiveness of ICs, Microsoft attempted to convert some of its employees to ICs. Their lawyers knew that it was improper because the workers did not qualify as ICs.

The IRS questioned the conversions and Microsoft admitted that they were really employees. The IRS reclassified them. Upon hearing of Microsoft's IRS admission, the reclassified employees got together and sued Microsoft, claiming that they were entitled to the employee benefits that other Microsoft employees received. Microsoft had to pay them a large amount of money. The moral is not that IC conversions are improper; rather, the moral is that it has to be properly done and for the right reasons.

Microsoft's lawyers did them a great deal of damage because they didn't provide their client with the correct advice. The Microsoft case is not a reason for you to fear an IC conversion.

Allstate was recently sued by a group of salesmen, who claim they were "coerced" into becoming ICs. Well, it should be obvious that coercion should not be involved. If the worker does not want to be an IC and obtain the many benefits, he is obviously not a good candidate for IC treatment.

CHAPTER 8: PERSPECTIVE OF A BUSINESSMAN

The fact that you are considering using IC's in your business should be only one aspect of an overall plan to make your business an efficient and profitable operation. It is not enough that your business is doing well, if none of the profit is going into your pockets. Unfortunately, so many business advisors and accountants do not understand this.

It is often difficult for a business owner to see the inefficiencies in his own business and, therefore, outside consulting is necessary. It is not the purpose of this Book, nor is it possible, to analyze all of the requirements for an efficient and profitable business because each business is unique. If you want a glimpse into the near future get the recent works of management and business guru, Peter Drucker.

However, we believe that the use of IC's is the first step to the organization of a sound business operation. It is not only a reduction of cost item; it is a structural change, which will make your business more efficient and will make your SPs happier and more independent.

As I am sure you are all aware, the term outsourcing is relatively new. Yet, it signifies what is happening. Companies are outsourcing all extraneous practices, so that they can stress their core competencies. The movement toward the use of IC's is just a small part of the outsourcing movement. Employee leasing and contingent workforces, however, are more expensive and have many more drawbacks than the use of ICs.

Our employment structure is undergoing massive change. The new technologies are changing it. We are quickly becoming one world. Geographic boundaries are and will be quickly evaporating. Information control is no longer possible. The old power bases are dissolving. Income and sales taxes. as they presently exist, are no longer viable. Social Security is a highly visible and unresolved problem.

There is no such thing any longer as a lifetime job. Changes are averaging every five years. Pensions, health and life insurance need to be owned by the individual worker. Companies can and do go bankrupt.

This is the concept of the free market that allows new businesses to thrive.

Information and knowledge belong to the individual, not to the company. Knowledge can be copied and it is mobile. The new knowledge worker—the Business of One—has provided us with the greatest increase in productivity in the history of the world. Businesses that operate the old way are doomed to failure.

The outstanding feature of America's continuing economic miracle is freedom of choice, despite the tendency of government to limit freedom. We Americans always seem to find a way around the restrictions thrown at us by the politicians and bureaucrats.

The key to building a successful business in this new age is to build alliances not adversaries and the people who work for you are your allies and they need to be kept happy. Build your business not your overhead.

Make yourself into a virtual business. The advantages of an IC operation have already been explained to you. Take the plunge and free yourself. Accountants and lawyers are basically nitpickers. They are good at undoing things but poor at devising new ways of doing things. Why is it that lawyers, doctors, court reporters, most accountants, etc. are considered ICs? Do they know something that we don't?

CHAPTER 9: CONVERSIONS

In order to convert present employees to ICs a special provision of the SHT requires that the employee's new position with your company not be substantially the same that he held as an employee. The SHT cannot be invoked for protection, if your converted employee was in a "substantially similar position" before he was converted to an IC. This does not mean that he does not qualify under the 20CLFT because the 20 CLFT is separate from the SHT. But it does mean that you will have to prove the IC status under that test. We have already shown you how to do this.

But it is always better to have the protection of the SHT. Therefore, what do you have to do to qualify your converted employees? Well, what does "substantially similar position" really mean? I believe it means that you must do more than just calling a person an IC, as did Microsoft.

It means a restructuring of the work that the employee did. Is he now a supervisor, a consultant, a manager, a specialist, etc? As we have stated, the use of IC's is more than a cost saving to you. It is a different way of looking at work. Give the worker a task and let him do it his way. As long as the end product is satisfactory why would you want to control him? He can rent the necessary tools from you or purchase them from you and do the job on his own.

The job that the employee now does for you has to be of a type that lends itself to the work of an IC. This requires a change of thinking. Why do you care when your IC comes to work? If you do, then forget about IC treatment. Why do care where he does the work? If you do, then keep him as an employee. Although if it is necessary for the work to be done on site, that is okay now under the new IRS rules.

Why do you care if he only works for you? If you do, then keep him as an employee, although the IRS is no longer interested in whether there are multiple sources of work.

You should be interested in allowing your SP to make as much money as he desires and if you don't have enough work for him, let him get other work. If there are secrets you don't want him to share with other companies, that can be taken care of with a proper secrecy clause in the IC Agreement.

If the employee doesn't understand the advantages you are providing by allowing him to work as an IC, then don't try to convert him to an IC. He won't hack it. And coercion can't be involved, unless you want to buy a lawsuit, as did Allstate.

CHAPTER 10: ACCOUNTING AND TAX SAVING TIPS

The independent contractor operation is a model of accounting efficiency and tax savings. The SR is probably more attuned to accounting needs than is the SP because, as a business owner, you have been required to keep books. The SP may or may not be familiar with accounting practices and the tax accounting aspect of running his business. These are some tips.

You are no longer bound to do payroll, except for those left as employees, and it may be wise to maintain a few. It is suggested that some of the staff should still be considered employees and payroll done for them, but it is not absolutely necessary, except for payments you may receive as a corporate officer, if you own a corporation.

It is no longer necessary to make withholding bank deposits for the IC's—payments to SP's should be in accordance with invoices received from the SP. The only forms necessary for IC's is the 1096 and 1099s for each of the IC's. An accountant is not necessary to complete these simple forms. Quickbooks software is recommended for businesses.

No matching Social Security Medicare for IC's. Cost saving of .0765 % of payroll.

Unemployment (FUTA) costs are eliminated—varies between 2.7%-10% of payroll.

Fringe benefits can be eliminated to the extent desired or kept, as the case may be.

The intangible cost saving of all the paperwork and legal restrictions attached to the employer-employee relationship and the tangible cost of workman's compensation insurance savings.

The following are some of the tips that you can advise your SPs that will be available to them. If the SP understands the benefits of an IC relationship, he is not apt to claim "coercion" or to file lawsuits. The relationship is subject to arms length negotiations.

Schedule C entitlement for deductions of business related expenses from gross income results in a substantial reduction in taxable income. See IRS form Schedule C.

Even though Social Security and Medicare are taxable at the full rate (15%), Form 1040, page 1 allows a deduction for one-half; therefore the Social Security burden is no more than it would be for an employee paying half and employer paying half. Furthermore, Social Security tax is taken only on the net income after deduction of expenses.

Home Office expense are deductible as a percentage of total house use or as a reasonable lease from owner to IC. Home office equipment, utilities and furniture are deductible either by depreciation or lease.

All the Internal (Infernal) Revenue Code goodies and benefits are available to IC's, as well as to the wealthy, using a wise financial/tax advisor, who can take advantage of the tax breaks that are scattered throughout the Code.

Car and Truck expenses are deductible based on actual cost at a percentage of usage or mileage can be used, if documented, with an easy to keep trip calendar. Mileage begins from the home office.

Business equipment is fully deductible (through depreciation schedules or as a lease) against gross income. Your SPs should have a computer and one of the popular accounting software programs.

Pension plans— HR 10 (Keough Plan)—Only available to IC's, as self-employed individuals, up to 25% of gross income for a maximum of $30,000.00, last time I looked at it. This plan is available even though the SP or his spouse are participants in another employee plan.

S.E.P. (Simplified Employee Plan)—an IC is considered an employer and employee contributions can be up to 15% of net income.

IRA. (Individual retirement plan)—maximum $2,000 deduction (subject to gross income limitations) and the Roth IRA and 401 k's.

Simple Plan—subject to IRA. limitations.

If the SP is a corporation and he is a corporate officer, he can take income both as an employee and as self-employed and wear two hats and, therefore, obtain the tax and pension advantages of both an IC and as an employee. See, *infra*.

Insurance—there are two perspectives on this subject. One is from the view of the SP and the other from the SR. The SP will in all likelihood need a greater input on this subject because he may be unfamiliar with it. Of course a lot depends on what coverage the SR has in place and what he intends on keeping. Legally, there is an elimination of workman's compensation (which should be covered by the SP with disability insurance).

Also, there is no unemployment compensation. This can be covered by an adequate severance provision in the IC agreement and adequate savings for emergencies by the SP.

The SR may continue health and dental insurance or another group plan can be arranged. General liability ("umbrella") or Errors and Omissions insurance may also be necessary.

Pension and investment advice is needed for both the Recipient and the Provider, especially where the Recipient is a corporation and the officers opt to be IC's to the corporation and, therefore, eligible for the great benefits of the Keogh and other plans.

PART IV
FOR CORPORATE OFFICERS

CHAPTER 11: DUAL HAT STATUS

We assume you are already incorporated and are an officer of that corporation (President, CEO, Vice-President, Treasurer, Secretary etc.) because dual hat status only applies if you hold that status. But remember, an SP can be an employee and an IC.

The IRS has ruled that Corporate Officers, previously thought of as being exclusively statutory employees and, therefore, subject to employment status, and all of its concomitant laws, rules and regulations can wear a "dual hat" and be both an employee and an IC. Your corporation is your SR and you are the SP/Employee. You have to meet the easy but specific requirements of what constitutes an independent contractor, which has to be separate from your role as a corporate officer employee.

What difference does dual status make? As an employee, you can participate in your corporation's qualified retirement plan and be eligible for fringe benefits. Getting those can be a big plus.

As an IC you can set up a Keogh plan and tax shelter additional amounts for ultimate retirement.

You can also deduct business expenses that would, if related to employment as an employee, be deductible, if at all, as a miscellaneous itemized deduction. Such deductions are limited, as they are subject to a 2% floor and a 3% phase out.

An additional plus that is affecting more and more taxpayers is that miscellaneous itemized deductions are not deductible at al l for purposes of the Alternative Minimum Tax. Although these benefits may seem complicated, they aren't. And dual hat status will save you a pile of money that would otherwise go to Uncle Sam. Best to talk to a trustworthy counselor about all the tax angles.

Besides the retirement and tax benefits, already mentioned, there are others. For instance, as an IC you can get others to pay you for the services you provide to your corporation.

The task of dual hat status is to avoid interrelatedness. Your function toward your corporation must be different than the function you perform as a corporate officer. They must not be related. If you receive pay as the corporate officer, this must be reported as withholding wages. But the IC pay will be as a self-employed individual.

In its IRS legal memorandum (ILM) 200038045), prepared by the IRS's Chief Counsel's Office on August 9, 2000, the IRS opined that, although a corporate officer was a "statutory employee," there was no exclusion in the SHT that prevented a corporate officer from also being an IC (SP) to his corporation (SR). Such an individual could wear "two hats."

This IRS Chief Counsel's opinion, which is almost like law, provides corporate officers with a tremendous advantage over exclusive employee status.

This was a great, although mostly unnoticed and unused benefit to owners/officers of corporations. As stated by the author of the ILM, "…submission of a Form 1099 may be sufficient to establish a corporation's treatment of the officer as an independent contractor."

As I have stated, lack of interrelatedness is the essence of dual hat status. There are many ways of doing this. You can be an officer of your corporation, who does the bookkeeping. but, also, be a manager of your sales force or a salesman, or a board member, or a carpenter, etc—the list is without limit. The employee income part of the dual hat will be reported on Form W2 and the IC part on Form 1099.

CHAPTER 12: SETTING UP UNDER THE "TWO HAT SYSTEM"

Set up a corporation.

First Hat—**Pay yourself a reasonable salary as a corporate officer.**

Second Hat—**Set up an IC Agreement with your corporation, which will pay you from the net profits for services rendered. Payments to be reported by the corporation to the IRS with form 1099. Payments to you are reflected on 1040 Schedule C. 25% of the gross Schedule C income but not more than $30,000 can be sheltered in a Keough pension plan, which you control.**

Additionally, take this as an EXAMPLE of what can be done:

You have been in a service business for a number of years. Your wife formed a Corporation to perform her own service business. You sell services to her corporation, pursuant to an IC Agreement. You already have health insurance. However, pursuant to the IC Agreement, your wife's corporation can pay for that.

You have a cost efficient automobile, and can pay for the expenses using the standard mileage option as a Schedule C deduction. This will be beneficial both to the corporation (SR) and to the IC (SP).

This arrangement will allow you to expense a portion of household expenses for use as a "second office" at your home. You also have the option of starting your own retirement plan (known as a Keough or HR10). This fantastic plan is under your control, unlike 401k's. And the deductible tax-free amount you can contribute to it greatly reduces taxable income by up to 25% of gross Schedule C income up to $30,000.

Also, you are shielded from an IRS audit that might deem your pay excessive, if you were paid as a corporate officer, and charge it as "disguised dividends," which are not deductible by the corporation. These disguised dividend adjustments do not apply to IC's. You can also rent your equipment and car from the corporation.

WHY AND HOW

The benefits to your wife's corporation (SR) are as follows:

1. None of the complexities and expenses of payroll reporting.

2. No payroll taxes and no accounting fees to keep track of it. Just a simple 1099 at the end of the year.

3. Simple reimbursement for medical expenses;

4. Corporation doesn't have to provide you with a car.

5. You can dispense your services pursuant to the IC Agreement.

6. Corporation doesn't have to pay for Worker's Compensation Insurance.

7. Corporation won't be burdened with having to comply with all the federal and state laws applicable to employees, a costly burden in terms of time necessary to complete all the forms.

8. Corporation doesn't have to be concerned about being sued for any of the number of lawsuits permitted by employees that have so burdened the employment system (ES).

In exchange for the above listed corporate advantages, you as the self-employed SP can negotiate a higher percentage of billable fees and work for other clients, plus obtaining the tax advantages previously mentioned. Depending on the terms of the IC Agreement, it is estimated that a corporation (SR) can increase payments to you by between 25%- 40% because of the cost savings. A cost analysis that will provide a detail of the corporation's cost savings of switching to a dual hat operation can be easily done by a good accountant.

APPENDIX

LAW

There is so much law on the subject of independent contractors that a whole warehouse would be necessary to contain it. The following is the best of the best.

Smoky Mountain Secrets, Inc. v. U.S., 910 F. Supp. 1316 (E.D. Tenn.,1995) is what lawyers call a seminal case on the IC issue. It provides the protection (through the use of IC Specialists) to guaranty that you are acting properly under the law. The following excerpts show how far we have come in the battle to win work freedom. The following is from the Court's opinion:

...Mr. Gee is a CPA who has been licensed and in practice for more than 20 years. He received his undergraduate degree in accounting from Western Kentucky University and Masters in Business Administration from the University of Tennessee at Knoxville.
* * *
...Mr. Gee opined that SMS's telemarketers and delivery persons were direct sellers as contemplated by (the law). Mr. Gee testified that his opinion was based upon (his client's) description of the relevant facts about SMS's business, the manner in which the telemarketers and delivery persons would be compensated, and the fact that plaintiff had a written contract with its sales force providing that the telemarketers and delivery persons would not be treated as employees for federal tax purposes.
* * *
Like Mr. Gee, Mr. Sharpe (a CPA) advised (his client) that he believed that SMS's telemarketers and delivery persons were properly classified as independent contractors and not employees. He based this advice, however, upon his prior experience and knowledge of common law (20 common law factors).
* * *

WHY AND HOW

After again analyzing SMS's sales force to the context of these 20 factors, Mr. Sharpe advised (his client) that the telemarketers and delivery persons were properly characterized as independent contractors.

* * *

Thus, the only remaining question is whether SMS had a reasonable basis for treating its telemarketers and delivery persons as independent contractors. The term 'reasonable basis' is to be construed liberally in favor of the taxpayer. (Citing authority).

* * *

SMS claims that its reliance on the advice of two advisors is sufficient to demonstrate a reasonable basis under Sec.530 (the Safe Harbor) for not treating its telemarketers and delivery persons as employees. I agree. Under the circumstances of this case, reliance upon the professional advice rendered by two CPA's…constitutes a reasonable basis for SMS having treated its telemarketers and delivery persons as independent contractors.

* * *

I further conclude that SMS's reliance upon the advice of Mr. Sharpe, who examined the information provided by (his client) in the context of the common law factors governing independent contractor status, was reasonable, thereby entitling SMS to the protection of Sec. 530.

* * *

Generally, the courts have found that reasonable cause exists where the taxpayer relied on the advice of a trusted attorney or accountant. (Citing authority). Indeed in this regard, the Supreme Court has stated that:

'When an accountant or attorney advises a taxpayer on a matter of tax law, such as whether a liability exists, it is reasonable for the taxpayer to rely on that advice. Most taxpayers are not competent to discern error in the substantive advice of an accountant or attorney. To require the taxpayer to challenge the attorney, to seek a "second opinion," or try to monitor counsel on the provisions of the Code himself would nullify the very purpose of seeking advice of a presumed expert in the first place. "Ordinary business care and prudence" do not demand such actions." (Citing Supreme Court case).

BIBLIOGRAPHY AND WEBSITES

Jimmy Moore, *Advancing Into Temp, Contract, and Consulting Jobs* (Writer's Club Press an Imprint of Iuniverse.com (2001).

Robert W. Wood, *Legal Guide to Independent Contractor Status*, (Third Edition, A Panel Publication, Aspen Publishers (2000).

Willie Jackson and Edgar H. Gee, Jr. and Michael J. Knight, *How to Shift the Burden of Proof to the IRS on Independent Contractor Status*, 28 Tax Advisor No. 10, 642 (1997)

Russell A. Hollrah, *Employer's Handbook, Independent Contractor* (Thompson Publishing Group).

The IRS website has a wealth of information and, if you feel that you need it. Of course it is mostly confusing. There is enough blah, blah, blah to fill your house. I have simplified all the complicated language to the bone. However, if you feel the need, go to www.irs.gov and search the words, "independent contractor."

There are some necessary forms that can be obtained at the IRS site.

• IRS Form 1099. This is the only form that your SR will provide to the IRS and you, the SP. You may need to file estimated payments. See your tax a dvisor about this. The 1096 Form (for SR's) accompanies the various 1099's and is just a listing of all the 1099 Forms that you may file.

• SS4-Application for EIN. This is an application for an Employer's Identification Number. It should be used by SR's or SP's who want to start a business. It is necessary to open a bank account, which should be separate from your personal account.

• SS9-Request for Taxpayer ID. The SP's should be asked by your SR to complete and return. It prevents the SR from being penalized by the IRS.

Also, the many websites that claim to cater to Independent Contractor businesses shows the popularity of independent contractor status. However, they provide complicated instructions and unnecessary scare tactics. A simple search of your favorite search engine under "independent contractor" will provide you with more hits than you can handle. The best of the sites are www.nolo.com (which also provides good how to do it information on choosing a business entity and other small business topics). Also, Attorney Urquhart's website www.workerstatus.com has a wealth of material.

IC QUESTIONNAIRE

This is provided only as an example of information you might want to gather from your workers. Of course, each business is different and you need to determine the qualifications you need.

FOR CONSTRUCTION SERVICE PROVIDER
DATE
NAME
BUSINESS NAME CONTACT
BUSINESS ADDRESS
PHONE FAX MOBILE PAGER
HOME ADDRESS
SS# FEDTAX#
REFERRED BY WHOM
ARE YOU PRESENTLY DOING CONTRACTING JOBS?
HOW MANY HELPERS DO YOU NORMALLY USE?
DO YOU HAVE ACCESS TO ADDITIONAL HELP?
IS SOMEONE AVAILABLE TO ANSWER YOUR PHONE?
DO YOU HAVE WORKER'S COMPENSATION INSURANCE?
DO YOU HAVE GENERAL LIABILITY INSURANCE?
DO YOU HAVE TRUCK/AUTO INSURANCE?
WILL YOU DO EMERGENCY WORK OUT OF REGULAR
BUSINESS HOURS?
WILL YOU DO REGULAR WORK AT NIGHT OR WEEKENDS?
HOW FAR WILL YOU TRAVEL FOR WORK?
TRADE OWN OR RENT SKILL LEVEL
SPECIALTY EQUIPMENT
DEMOLITION
DEBRIS REMOVAL
CLEANING-STRUCTURE / -SOFT / FURNITURE
CLEANING—CARPET
ROUGH CARPENTRY / FINISH CARPENTRY
DRYWALL / STUCCO / PLASTER
CABINETS / WINDOWS
ACOUSTIC TILE
PAINT INTERIOR / PAINT EXTERIOR

STRIP AND FINISH / SAND AND FINISH
WALLPAPER
INSULATION
CERAMIC TILE
CARPET INSTALLATION
VINYL INSTALLATION
WOOD FLOOR INSTALLATION
ROOFING
MASONRY
VINYL/ALUM. SIDING/TRIM
GARAGE DOORS
ELECTRICAL
PLUMBING
HEATING/AC

This should be a checklist that needs to be utilized for each new hire.

IC INFORMATION FORM

TO:
FROM: DATE:
IN ORDER FOR US TO UTILIZE YOUR SERVICES AS AN
INDEPENDENT CONTRACTOR WE NEED THE FOLLOWING
INFORMATION:
___W-9 FORM (FILL OUT, SIGN AND RETURN)
___COPY OF YOUR LIABILITY INSURANCE COMPANY
___COPY OF YOUR DRIVERS' LICENSE
___IF YOU EMPLOY ANYONE, A COPY OF YOUR
    WORKMAN'S COMPENSATION INSURANCE
___BUSINESS CARD
___RECENT ADS IN NEWSPAPER
___SOCIAL SECURITY NUMBER OR FEDERAL I.D. NUMBER
___COPY OF YOUR ARTICLES OF INCORP. OR DBA CERT.
___INDEPENDENT CONTRACTOR QUESTIONNAIRE
___OTHER.
___TWO REFERENCES YOU RECENTLY WORKED FOR.
    NAME - PHONE -   / NAME - PHONE -

INDEPENDENT CONTRACTOR

This is a pre-hire checklist used by a noted university. It is provided only as an example that can be adapted to your needs.

IC PRE-HIRE WORKSHEET USED BY A UNIVERSITY

Individual /Sole Proprietor/ Corporation
Social Security Number_____
Federal ID Number_____
Name_____
Name of Company_____
Campus_____
Department_____
If Foreign National—Country_____
Visa Type_____

MULTIPLE RELATIONSHIPS WITH THE UNIVERSITY

1. Is this individual on record as a current employee?
Yes   No
If no, is it expected that the University will hire this individual as an employee following the termination of this service?
Yes   No
2. Was the individual a University employee any time during the last year and did he or she provide the same or similar services while an employee?
Yes   No

IRS CLASSIFICATION FACTORS

Before a worker is hired as an independent contractor, the following checklist *must* be completed to help determine whether an employer/ employee relationship exists.
IRS Classification Factors Yes =
Employee No =
Contractor Behavioral Control: Right to direct and control details and means by which worker performs services.
1. Instructions. Will the University have the right to give the worker instructions about when, where, and how he or she is to do the job?

2 Training. Will the worker receive training from the University? Financial Control: Right to direct and control economic aspects of the worker's activities.

3 Significant Investment. Has the worker failed to invest in facilities (such as an office) used to perform services?

4 Payment of Expenses. Will the University pay the worker's business or travel expenses?

5 Services Available. Does the worker not make his or her services available to other employers?

6 Payment by Hour, Week, Month. Will the University pay the worker by the hour, week, or month rather than by commission or by the job?

7 Realization of Profit or Loss. Will the arrangement prevent the worker from realizing a profit or suffering a loss?

Relationship of Parties: Intent of parties concerning status and control of worker.

8 Written Contract. Will a written contract not be executed describing the worker as an independent contractor?

9 Employee Benefits. Will the worker receive any employee benefits?

10 Right to Terminate. Could the University terminate the worker at any time without incurring liability?

11. Regular Business Activity. Is the work to be performed part of the regular business of the University, such as teaching or research?

EVALUATION OF CLASSIFICATION FACTORS
Areas That Support Employee Status
Areas That Support Contractor Status (Use separate sheet, if necessary.)
DETERMINATION
Hire worker as an employee
Hire worker as an independent contractor

_____

Department Authorization
Prepared By_____
Date_____

SP INSTRUCTIONS FOR COMPLETING AN IC AGREEMENT

The following details what the SP should consider when signing an IC Agreement:

Please provide the following information:

Home addresses.

Social Security #s.

The function(s) each of you performs for the corporation, such as sales, field superintendent, marketing services, office management, etc.

The rate of pay you intend to receive, whether hourly, daily, weekly, biweekly, monthly, piecework, or accomplishment of certain tasks.

The terms of the agreement.

How it will work:

You will submit an invoice in the following form to the Service Recipient.

(YOUR LETTERHEAD)

INVOICE #

FOR SERVICES RENDERED FROM (DATE) TO (DATE)

TOTAL AMOUNT OWED $

This is an example of an ad that can be put up on a grocery store bulletin board:

(Name)

(Telephone)

Accepting jobs for Services (describe)

Please call for appointment—tel. #

INSTRUCTIONS FOR SR'S

Obtain a signed completed checklist for IC's from the SP.

The SP is a task oriented individual; he is restricted by you only to the extent of the proper performance of that task. The hours of his work should not be controlled by you but should relate to the time it takes to do the task. You may provide general instructions as to what the task is but you should not instruct the SP with specific instructions on how to do the task.

If the person you are hiring is already one of your employees, you should define the new tasks you will be asking him to do that is different than what he is now doing. Usually attaching the management function to the task presently being performed should suffice.

Pay should be determined after obtaining an analysis of labor costs. Pay is to be negotiated, taking also into consideration the cost of any benefits you want to provide. The entire IC agreement can be negotiated as you and the SP may agree. Pay can be by piece work (the items involved in the task), hourly, weekly, etc. Provision should be made to include discretionary (at your discretion) bonuses (either year end or other) but this is optional.

You do want to keep people who work for you happy and you must understand that your savings need to be passed down to your IC's, who will have to pay for a lot that you used to pay. Payment is to be made only upon receipt of the SP's invoice.

After all parties sign the IC agreement, the SP should receive a copy and the original placed in his personnel file.

The only IRS documents necessary for filing are the 1099 year end statements (due by January 31st of each year) and the 1096 summary statement (which is a summary of all the 1099's you file). The SP is given a copy for him to use in completing his tax forms.

Besides whatever costs savings you obtain, you should also benefit from increased productivity and a more pleasurable workplace and workforce.

OUTLINE FOR ESTABLISHMENT OF AN IC RELATIONSHIP

Before interview
Advertise for contractors or service providers.
Use appropriate application form.
During interview—make prospective worker feel good about the IC arrangement.

INDEPENDENT CONTRACTOR

1. Use proper terms (contractor, service provider, representative) vs. employee, job, and salesman.

2. Assume that person knows the difference between an employee and an independent contractor and discuss it from that point of view.

3. Explain that all your workers are contractors (service providers) and they want it that way because the advantages outweigh the disadvantages, especially gross pay and allowance of business deductions—(list others from outline). Allowable expenses are often permitted as a percentage of gross and usually amount to 50% of gross or more. The amount is different for each industry but accountants are aware of the percentage. At end of year a 1099 form is sent just like a bank does for interest.

4. Payments are made in full with no deductions and can be structured to take care of estimated tax payments, if contractor chooses to make estimated payments. Half of self-employment taxes are now deductible and it is probable in the future that they may be fully deductible.

5. You will assist at end of year to do taxes, if this is new. You have accountants who do other contractors. Also, you will provide assistance in finding appropriate insurance agents for liability, health, disability and pension information.

GENERAL IC AGREEMENT

No form independent contractor agreement can particularize your special needs. The following Agreement is a general one that can be adapted to your needs:

AGREEMENT IS HEREBY MADE between the SERVICE RECIPIENT and SERVICE PROVIDER set forth below according to the following terms, conditions, and provisions:

1. IDENTITY OF SERVICE RECIPIENT. Service Recipient is identified as follows:
Name:
Type Entity: ( ) Sole Proprietorship ( )Partnership
( ) Corporation ( )Limited Liability Company
( ) d/b/a ( ) Other
Address:
City/State/Zip:

Business—Tele.: Fax:

2. IDENTITY OF SERVICE PROVIDER. The Service Provider is identified as follows:

Name:

Type Entity: ( ) Sole Proprietorship ( )Partnership

( ) Corporation ( )Limited Liability Company

( ) d/b/a ( )Other

Address:

City/State/Zip:

Business—Tele.: Fax:

Social Security or Federal I.D. #

3. NATURE OF RELATIONSHIP. The Service Provider is an independent contractor in his relationship with Service Recipient. This means, as follows:

a. Service Recipient has neither the right to nor shall he exercise any control or direction over the methods by which Service Provider performs the duties and functions of Service Provider. Service Provider will devote his best efforts in the performance of his duties as hereafter set forth;

b. Service Provider is an independent business person who offers his services to the public at large and has the right to perform services for others during the term of this agreement. Service Provider has the sole right to control and direct the means, manner and method by which the services required by this agreement will be performed. Service Provider has the right to perform the services required by this agreement at any agreeable location and the time spent by Service Provider is solely at his/her discretion within any monetary caps which are agreeable to both parties;

c. Service Provider owns or rents the tools used in performance of his services to Service Recipient;

d. Service Provider is responsible for payment of his own taxes and shall make no claim against Service Recipient for pension benefits, liability insurance, or medical insurance, other than as may hereafter be stated in this contract;

e. Service Provider is responsible for his own negligence and hereby specifically holds Service Recipient harmless and agrees to defend and indemnify Service Recipient.

f. Service Recipient shall not be liable to Service Provider for any expenses paid or incurred by Service Provider, unless otherwise agreed in writing.

g. Service Provider shall supply, at Service Provider's sole expense, all equipment, tools, materials, and/or supplies to accomplish the job agreed to be performed.

h. Neither federal, nor state, nor local income tax, nor payroll tax of any kind shall be withheld or paid by Service Recipient on behalf of Service Provider or the employees of Service Provider. Service Provider shall not be treated as an employee with respect to the services performed hereunder for federal or state tax purposes.

i. Service Provider understands that Service Provider is responsible to pay, according to law, Service Provider's income tax. If Service Provider is not a corporation, Service Provider further understands that Service Provider may be liable for self-employment (social security) tax, to be paid by Service Provider according to law.

j. Because Service Provider is engaged in Service Provider's own independently established business, Service Provider is not eligible for, and shall not participate in, any employee pension, health, or other fringe benefit plan, of the Service Recipient.

k. No workers' compensation insurance shall be obtained by Service Recipient concerning Service Provider or the employees of Service Provider. Service Provider shall comply with the workers' compensation law concerning Service Provider and the employees of Service Provider .

l. Service Recipient is not responsible for Unemployment Compensation and will not pay any Federal Unemployment Tax on behalf of Service Provider and Service Provider expressly waives and releases Service Recipient for any responsibility for same and Service Provider hereby states that no claim for such benefits will be made.

m. Service Provider warrants that he has entered into this agreement with full knowledge and understanding of its terms freely and voluntarily.

4. PROJECTS TO BE PERFORMED. Service Recipient desires that Service Provider perform and Service Provider agrees to perform the following project(s):

5. TERMS OF PAYMENT. Service Recipient will pay a fee of $ (Terms of Payments)
Service Recipient shall pay Service Provider 's fee within a

reasonable time after receiving Service Provider 's invoice. Invoices shall be submitted on Service Provider's letterhead specifying: (i) an invoice number, (ii) the dates covered in the invoice, (iii) an outline of the work performed during the period.

6. TERM OF AGREEMENT. The period within which the Service Provider services are to be rendered under this Agreement shall commence on   and shall, at the latest terminate on   , but may be renewed for the same additional term if agreed to by both parties.

7. TERMINATION WITHOUT CAUSE. Without cause, either party may terminate this agreement after giving 15 days prior written notice to the other of intent to terminate without cause. The parties shall deal with each other in good faith during the 15-day period after any notice of intent to terminate without cause has been given.

8. TERMINATION WITH CAUSE. With reasonable cause, either party may terminate this agreement effective immediately upon the giving of notice of termination for cause. Reasonable cause shall include:

A. Material violation of this agreement.

B. Any act exposing the other party to liability to others for personal Injury or property damage.

C. Unacceptable quality of work.

9. NON-WAIVER. The failure of either party to exercise any of its rights under this agreement for a breach thereof shall not be deemed to be a waiver of such rights or a waiver of any subsequent breach.

10. NO AUTHORITY TO BIND SERVICE RECIPIENT . Service Provider has no authority to enter into contracts or agreements on behalf of Service Recipient. This agreement does not create a partnership or agency relationship between the parties.

11. DECLARATION BY SERVICE PROVIDER. Service Provider declares that Service Provider has complied with all federal, state and local laws regarding business permits, certificates and licenses that may be required to carry out the work to be performed under this agreement.

12. HOW NOTICES SHALL BE GIVEN. Any written notice given in connection with this agreement shall be given in writing and shall be delivered either by hand or by certified mail, return receipt requested, to the party at the party's address stated herein. Any party may change its address stated herein by giving notice of the change in accordance with this paragraph.

13. ASSIGNABILITY. This agreement may not be assigned, in whole or in part, by Service Provider.

14. CHOICE OF LAW. Any dispute under this agreement or related to this agreement shall be decided in accordance with the laws of the State of   .

15. ARBITRATION. Any dispute which may arise between the parties shall be submitted to binding arbitration. Each party shall select an Arbitrator and the two Arbitrators shall select a third. The decision of the Arbitrators shall be binding and final and shall be subject to enforcement in a Court action. Prior to any arbitration decision, costs of arbitration shall be divided equally by the parties.

16. ENTIRE AGREEMENT. This is the entire agreement of the parties.

17. SEVERABILITY. If any part of this agreement shall be held unenforceable, the rest of this agreement will nevertheless remain in full force and effect.

18. AMENDMENTS. This agreement may be supplemented, amended or revised only in writing by agreement of the parties.

19. ATTORNEY'S FEES. If any legal action or arbitration or other proceeding is brought for the enforcement of the Agreement, or because of an alleged dispute, breach, default, or misrepresentation in connection with any of the provisions of the Agreement, the successful or prevailing party shall be entitled to recover reasonable attorneys' fees and other costs incurred in that action or proceeding, in addition to any other relief to which they may be entitled.

AGREED:

Dated:

_____

, Service Recipient

_____

, Service Provider

THE IRS IS MOSTLY ON OUR SIDE

This Training Manual has been edited. It is mostly the usual confusing bureaucratic gobbledygook and mumbo jumbo and not worth reading, as what I have previously stated to be the way to setup an IC status is the simple version. However, it is provided to form a basis as to how the IRS views IC's. The entire document can be found on the IRS website.

DEPARTMENT OF THE TREASURY • INTERNAL REVENUE SERVICE INDEPENDENT CONTRACTOR OR EMPLOYEE?

TRAINING MATERIALS

THIS MATERIAL WAS DESIGNED SPECIFICALLY FOR TRAINING PURPOSES ONLY. UNDER NO CIRCUMSTANCES SHOULD THE CONTENTS BE USED OR CITED AS AUTHORITY FOR SETTING OR SUSTAINING A TECHNICAL POSITION.

Training 3320-102(10-96) / TPDS 84238I
October 30, 1996

FOREWORD

Examiners and other Internal Revenue Service (IRS) representatives are sometimes faced with the difficult task of making a determination of the classification of workers who provide products and services for others. The status of a worker as either an independent contractor or employee must be determined accurately to ensure that workers and businesses can anticipate and meet their tax responsibilities timely and accurately. Businesses decide whether to hire employees or independent contractors depending on individual needs, customer expectations, and worker availability. Either worker classification—independent contractor or employee—can be a valid and appropriate business choice.

The majority of classifications of workers are not challenged by the IRS. When they are, there is usually agreement between the IRS and the business after the facts and circumstances are jointly reviewed. Nonetheless, when the IRS determines there may be a need for reclassification to accurately reflect the relationship of the worker and the business, the legal standard for distinguishing between independent contractor and employee can be difficult to apply. Also, the importance of indicators that might help in applying the legal standard can change and should be reviewed from time to time.

This training addresses the application of section 530 of the Revenue Act of 1978. Section 530 can in certain circumstances relieve businesses of employment tax liability resulting from worker classification. This training provides you with the tools to make legally correct determinations of worker classifications. It also discusses facts that may indicate the existence of an independent contractor or an employer-employee relationship and guides you in determining which facts are most relevant under the common law standard. It emphasizes that relevant facts may change over time because business relationships and the work environment change overtime. In addition, it addresses how to determine whether workers are statutory employees.

IRS policy requires its employees to exercise strict impartiality in the conduct of their duties. Thus, you must approach the issue of worker classification in a fair and impartial manner and actively consider section 530 relief at the beginning of an examination. This includes furnishing taxpayers with a summary of section 530 at the beginning of an examination. Additionally, you may need to assist taxpayers in identifying facts which establish either worker classification.

This course has been developed to provide Employment Tax Specialists and Revenue Officer Examiners with the tools to make worker classifications. The lessons will cover a review of the issues, law, and examination techniques for making a correct determination; as well as a review of Section 530 relief.

INDEPENDENT CONTRACTOR OR EMPLOYEE:
DOES SECTION 530 APPLY?
INTRODUCTION

Section 530 provides businesses with relief from federal employment tax obligations if certain requirements are met. It terminates the business's, not the worker's, employment tax liability under Internal Revenue Code (IRC) Subtitle C (Federal Insurance Contributions Act(FICA) and Federal Unemployment Tax Act (FUTA) taxes, federal income tax withholding, and Railroad Retirement Tax Act taxes) and any interest or penalties attributable to the liability for employment taxes (Rev. Proc. 85-18, 1985-1 C.B. 518).

WHY AND HOW

Section 530(e)(3) of the Revenue Act of 1978, as amended by the Small Business Job Protection Act of 1996, clarifies that the first step in any case involving whether the business has the employment tax obligations of an employer with respect to workers is determining whether the business meets the requirements of section 530. If so, the business will not have an employment tax liability with respect to the workers at issue.

At the end of this lesson, you will be able to:

1. Explain the two consistency requirements that must be met for a business to obtain relief under section 530.

2. Explain the reasonable basis test that must be met for a business to obtain relief under section 530.

3. Explain the three safe havens under the reasonable basis test.

4. Determine whether relief is applicable in a particular situation.

INTRODUCTION

Overview of requirements.

The business must meet the following consistency and reasonable basis requirements before the relief provisions of section 530 apply:

Consistency Test

The business must meet both aspects of the consistency test by:

• filing all required Forms 1099 (reporting consistency)

• treating all workers in similar positions the same (substantive consistency)

Reasonable Basis Test

The business must reasonably rely on one of the following:

• prior audit safe haven

• judicial precedent safe haven

• industry practice safe haven

• other reasonable basis

Meeting the consistency and reasonable basis tests will give the business relief from employment taxes with respect to the workers whose status is in question.

INTRODUCTION

Historical background

Section 530 of the Revenue Act of 1978, as amended, is not part of the Internal Revenue Code (IRC). However, some publishers include its text after IRC section 3401(a). It was originally intended as an "interim" relief measure, but was extended indefinitely by the Tax Equity and Fiscal Responsibility Act of 1982.

Section 530 was amended by section 1706 of the Tax Reform Act of 1986 (1986-3, C.B. (Vol.1) 698). Section 530(d) denies relief for certain technically skilled workers who provide services under a three party situation. It will be discussed in detail later in this lesson. Section 530(e) was added by section 1122 of the Small Business Job Protection Act of 1996 (H.R. 3448). Section 530(e), which is generally effective after December 31, 1996, contains a number of provisions that affect conditions under which a business will be eligible for section 530 relief. It is discussed throughout this lesson.

INTRODUCTION
Service must consider section 530

It is not necessary for the business to claim section 530 relief for it to be applicable. In order to correctly determine tax liability, as required by the IRS mission, you must explore the applicability of section 530 even if the business does not raise the issue. In addition, a plain language summary of section 530 must be provided to the taxpayer at the beginning of an examination of worker classification

Time to claim section 530 relief

The section 530 analysis is, itself, fact intensive. You will identify the possible application of section 530 relief before beginning the development of the worker classification issue. The relief is available, however, throughout the examination or administrative (including appeals) process, as well as, any subsequent judicial proceeding.

Section 530 limits guidance

When Congress enacted section 530, the IRS was barred from issuing any regulations or revenue rulings pertaining to worker classification. As a result, the IRS cannot issue new revenue rulings or even modify existing revenue rulings to reflect new developments. At the same time, courts have been able to modify their applications of the common law standard in response to factual developments. As a result, courts may now look at the employee versus independent contractor issue somewhat differently—possibly making outstanding IRS revenue rulings outdated and in conflict with judicial decisions.

Section 530 imposes no prohibition on private letter rulings or technical advice memoranda. Also there is no prohibition on published guidance dealing with section 530 itself.

INTRODUCTION

Section 530 considered first.

Section 530 is a relief provision that should be considered as the first step in any case involving worker classification.

Change from prior policy

Considering section 530 first is a change from prior policy and results from the Small Business Job Protection Act of 1996. New section 530(e)(3) specifies that a worker does not have to be an employee of the business in order for relief to apply.

Additionally, the business need not concede or agree to the determination that the workers are employees in order for section 530 relief to be available.

Other tax consequences for workers

A business may be entitled to relief under section 530 but workers may find, through a determination letter or some other means, that they have been misclassified and are employees. However, section 530 relief does not extend to the worker. It does not convert a worker from the status of employee to the status of independent contractor.

As noted above, misclassified employees are liable for the employee share of FICA rather than for tax under the Self Employment Tax Contributions Act (SECA). Workers may have filed and paid their own employment tax. If the worker paid SECA, the worker may file a claim for refund for the difference between SECA tax and the employee share of FICA. *See*, Rev. Proc. 85-18, section 3.08; Treas. Reg. section 31.3102-1(c). There are other tax consequences for the worker as well.

Workers as employees generally cannot deduct unreimbursed business expenses above the line on Schedule C, but must deduct them, if at all, as miscellaneous itemized deductions on Schedule A, Form 1040, subject to the two-percent limitation of IRC section 67. This sometimes results in liability for the alternative minimum tax.

Further, the worker as an employee cannot adopt or maintain a self-employed retirement plan. Finally, certain benefits provided by the business to a worker as an employee may be excludable from income by the employee due to specific IRC exclusions provided only to employees (*e.g.*, employer provided accident and health insurance).

CONSISTENCY TEST: REPORTING CONSISTENCY

Information Returns: Filing information returns

The first requirement a business must meet to obtain relief under section 530 is timely filing of all required Forms 1099 with respect to the worker for the period, on a basis consistent with the business's treatment of the worker as not being an employee. This provision applies only "for the period." Rev. Proc. 85-18, section 3.03(B). That is, if a business in a subsequent year (ed. Note: Period means Quarter not year) files all required returns on a basis consistent with the treatment of the worker as not being an employee, then the business may qualify for section 530 relief for the subsequent period. If a business is not "required to file," relief will not be denied on the basis that the return was not filed.

CONSISTENCY TEST: REPORTING CONSISTENCY

Information Returns:

Rev. Rul. 81-224 Rev. Rul. 81-224, 1981-2 C.B. 197, addresses specific questions about timely filing of Forms 1099. It provides that:

• businesses that do not file timely Forms 1099 consistent with their treatment of the worker as an independent contractor, may not obtain relief under the provisions of section 530 for that worker in that year.

• businesses that mistakenly, in good faith, file the wrong type of Form 1099 do not lose section 530 eligibility.

Best source: IRS records

The best source for determining whether Forms 1099 were filed timely is internal IRS records. Service Centers maintain information on the Payer Master File which records the taxpayer's history of filing information returns. These transcripts can be requested internally. Recall that Form 1099, reporting payments of $600 or more, must generally be filed by the last day of February following the close of the year in which the payment for the services was made. However, businesses may apply for extensions of time to file information returns.

CONSISTENCY TEST: SUBSTANTIVE CONSISTENCY

Substantive Consistency required

You will recall from reading section 530 that its provisions do not apply if the business or a predecessor treated the worker, or any worker holding a substantially similar position, as an employee at any time after December 31, 1977. In other words, treatment of the class of workers must be consistent with the business' belief that they were independent contractors.

Substantially similar position

A substantially similar position exists if the job functions, duties, and responsibilities are substantially similar and the control and supervision of those duties and responsibilities are substantially similar. In addition, section 530(e)(6), added by the Small Business Job Protection Act, states that the determination of whether workers hold substantially similar positions requires consideration of the relationship between the taxpayers and those individuals.

This includes, but is not limited to, the degree of supervision and control. This statutory change appears to be designed to enable differences in managerial responsibilities and differences in reporting requirements to be taken into account, along with differences in job duties.

Presumably, the contractual relationship and t he provision of employee benefits are also entitled to some weight. The determination of what is substantially similar work rests on analysis of the facts. The day-to-day services that workers perform and the method by which they perform those services are relevant in determining whether workers treated as independent contractors hold substantially similar positions to workers treated as employees. Comparison of job functions is an important fact. Workers with significantly different, though overlapping, job functions are not substantially similar.

CONSISTENCY TEST: SUBSTANTIVE CONSISTENCY

Defining treatment

Rev. Proc. 85-18 provides examples of treatment consistent or inconsistent with payments to an independent contractor:

1. The withholding of federal income tax or FICA tax from a worker's wages is treatment of the worker as an employee, whether or not the tax is paid to the Government.

2. Filing a Form 940, 941, 942, 943, or W-2 with respect to a worker, whether or not tax was withheld from the worker, is treatment of the worker as an employee for that period. NOTE: Beginning in 1995, household employers report wages paid to household employees on their individual income tax returns using Schedule H rather than Form 942.

3. The filing of a delinquent or amended employment tax return for a

particular tax period is not treatment of the worker as an employee if the filing was a result of IRS compliance procedures. However, filing the returns for periods after the period under audit is "treatment" of the workers as employees for those later periods, regardless of the time at which the return was filed.

4. Neither the use of an IRC section 6020(b) return prepared by the IRS nor the signing of Form 2504 (Agreement to Assessment and Collection of Additional Tax and Acceptance of Over-Assessment) constitutes treatment.

Changing treatment of workers

If the business begins to treat misclassified workers as employees, relief is available under section 530 for the years it treated them as independent contractors, provided it meets both consistency requirements (reporting and substantive consistency) and reasonable basis for the years prior to the change in treatment. *See* Rev. Proc. 85-18, section 3.04. The Small Business Job Protection Act added this rule as section 530 (e)(5).

Dual status

Some workers perform services in two capacities. For example, a business's bookkeeper might be separately engaged to design and print an advertising brochure. The fact that the bookkeeper is treated as an employee with respect to the bookkeeping services does not preclude application of section 530 if it is determined that the bookkeeper is an employee, and not an independent contractor, with respect to the design and printing services.

REASONABLE BASIS TEST

Moving to the next step. The business must reasonably rely on one of the following ways to meet the reasonable basis test, as listed in Rev. Proc. 85-18:

REASONABLE BASIS TEST EXPLANATION

Judicial Precedent Safe Haven

Reasonable reliance on judicial precedent; published rulings; a technical advice memorandum, private letter ruling, or determination letter pertaining to the business.

Past Audit Safe Haven

Reasonable reliance on a past IRS audit of the business for employment tax purposes, if the audit began after December 31, 1996, and entailed consideration of, but no assessment attributable to the business's employment tax treatment of workers holding positions substantially similar to the position held by the worker whose status is at (NOTE: A business may continue to rely on any audit that began before January 1, 1997, even though the audit was not related to employment tax matters.)

Industry Practice Safe Haven

Reasonable reliance on a long-standing recognized practice of a significant segment of the industry in which the business is engaged. The practice need not be uniform throughout an entire industry.

Other Reasonable Basis

A business which fails to meet any of the three safe havens may nevertheless be entitled to relief, if the business can demonstrate, in some other manner, any reasonable basis for not treating the worker as an employee. (Ed.Note: A determination that the worker is a common law IC is a "reasonable basis.")

Liberal construction

The Conference Agreement on section 530 of the Revenue Act of 1978 explains Congress' intent that the reasonable basis requirement be construed liberally. Extract H.R. Rep. No. 1748, 95th Cong. 2nd Sess. 4 (1978), 1978-3 C.B. (Vol. 1) 629, 633.

Generally, the bill grants relief if a taxpayer had any reasonable basis for treating workers as other than employees. The committee intends that this reasonable basis requirement be construed liberally in favor of taxpayers..

The Congressional direction to liberally construe section 530 means that facts which indicate that the conditions of section 530 have been satisfied by a particular business are to be viewed liberally in favor of the business. Liberal construction does not mean that the conditions for obtaining section 530 relief should be discounted or ignored. Failures to satisfy one or more of the conditions for eligibility for section 530 relief are not cured by the requirement of liberal construction of the reasonable basis requirement. (Ed. Note: So in the view of the IRS liberal construction does not mean liberal construction. I don't think so.)

Remember that if the business establishes the existence of a safe haven, the business must show reliance on the safe haven. Section 530 requires that the reliance must be reasonable. You should explore with the business why it treated the workers as independent contractors. The business's stated reasons should be set forth in your work papers.

REASONABLE BASIS TEST—PRIOR AUDIT

We will discuss the second reasonable basis safe haven first because section 530 relief is most easily established by reliance on a prior audit. A business is treated as having reasonable basis if it relied on a prior audit.

REASONABLE BASIS TEST—JUDICIAL PRECEDENT

Judicial precedent

Another safe haven provided by section 530 is judicial precedent. To obtain relief under this section, the business must demonstrate reasonable reliance on a judicial precedent, a published ruling, technical advice relating to that business, or a letter ruling to that business.

Qualifying TAMS and PLRs

A technical advice memorandum (TAM) or a private letter ruling (PLR) addressing the employer-employee relationship can be used by the business to which it was issued for judicial precedent safe haven.

REASONABLE BASIS TEST—INDUSTRY PRACTICE

Industry practice

The safe haven most commonly argued, and the one which causes the most controversy between businesses and the Government, is industry practice. Section 530 states that the business can claim reasonable basis if it can show reasonable reliance on a long-standing recognized practice of a significant segment of the industry in which the business is engaged. It makes sense to begin by defining "industry" since this establishes the group of businesses to be analyzed.

Industry defined.

Geographic area

An industry generally consists of businesses located in the same geographic or metropolitan area which compete for the same customers. For example, the landscaping industry will generally consist of businesses within a single metropolitan area. However, if the area includes only one or a few businesses in the same industry, the geographic area may be extended to include contiguous areas in which there are other businesses competing for the same customers.

If businesses compete in regional or national markets, the geographic area may include the competitors in that region or throughout the United States. For example, the commercial film production industry competes in a national market.

Long-standing

Whether a practice is long-standing depends on facts and circumstances. However, as confirmed by section 530(c)(2)(C), a practice that has existed for 10 years or more should always be treated as long-standing. The business may use the industry practice safe haven even if it began to provide a product or service after 1978. Similarly, a taxpayer may use the industry practice safe haven even if the industry came into existence after 1978. The legislative history clarifies that the 10 year rule is a safe haven. However, a shorter period may be long-standing, depending on the facts and circumstances.

Significant segment

How prevalent must the practice be to constitute a significant segment and/or recognized practice? Until the Small Business Job Protection Act amended section 530, neither the statute nor the legislative history provided any additional guidance on the appropriate standard for "significant." The determination was made on the basis of facts and circumstances, and it was an issue that often presented difficult analytical issues.

Establishing industry practice

Independent contractor treatment often flows from the business's general knowledge of competition in the industry or from communications with competitors or business advisers knowledgeable about the industry. Seldom will the business have performed a formal survey of industry practice at the time treatment of workers as independent contractors began. The fact that a formal survey was not conducted when independent contractor treatment began is relevant to, but is not conclusive of, whether the business relied on industry practice.

Do not automatically reject as irrelevant or immaterial a survey performed at or near the time of the audit. Such a survey can be relevant in establishing a business's prima facie case. The fact that a current survey confirms longstanding industry practice can buttress other evidence that the business relied on industry practice during the relevant period.

Reasonable reliance

The reliance required to satisfy the industry practice safe haven must be reasonable. Defining "reasonable" is a difficult task, but you might ask yourself: Would a reasonably prudent business under similar circumstances have relied upon such evidence of industry practice to treat workers as independent contractors? The extent of the business's knowledge of industry practice, whether obtained through personal experience, a survey, or through an advisor is relevant in this regard. The reasonableness or unreasonableness of the reliance may turn on the source of the information from which the business derived knowledge of the industry practice.

OTHER REASONABLE BASIS
Other reasonable basis

A business that fails to meet any of these three safe havens may still be entitled to relief if it can demonstrate that it relied on some other reasonable basis for not treating a worker as an employee. The legislative history indicates that "reasonable basis" should be construed liberally in favor of the taxpayer. H.R. Rep. No. 1748.

Advice of accountant or attorney

Reliance on the advice of an attorney or accountant may constitute a reasonable basis. The court cases tend to require the business to present (1) evidence of the educational and experiential qualifications of the attorney or accountant, and (2) evidence that the attorney or accountant issued the advice after reviewing relevant facts furnished by the business. *See, In re McAtee,* 90-1 USTC par. 50,242 (N.D. Iowa 1990) *vacating In re McAtee,* 89-2 USTC par. 9,625 (Bankr. N.D. Iowa 1989); *Overeen,* 91-2 USTCpar. 50,459 (W.D. Okla. 1991); and *Smokey Mountain Secrets, Inc. v. United States,* 76 AFTR 2d par. 95-5509 (1995).

The business need not independently investigate the credentials of the attorney or accountant to determine whether such advisor has any specialized experience in the employment tax area. However, the business should establish at a minimum, that it reasonably believed the attorney or accountant to be familiar with business tax issues and that the advice was based on sufficient relevant facts furnished by the business to the adviser.

State and non-tax federal law and determinations

Prior state administrative action (*e.g.*, workers' compensation decisions) and other federal determinations (*e.g.*, determinations under the Federal Labor Standards Act (Wage and Hour Division)) may or may not constitute a reasonable basis. This will depend on whether they use the same common law rules that apply for federal employment tax purposes. If the state or federal agency uses the same common law standard and interprets it similarly, however its determination should constitute a reasonable basis. If the state or federal agency uses a different statutory standard or interprets the common law standard differently, its determinations should not constitute a reasonable basis.

• *Queensgate Dental Family Practice, Inc., v. United States*, 91-2 USTC No. 50,536 (M.D. Pa. 1991)—The business treated licensed dentists as independent contractors based on the conclusion by the State Dental Board that state law prohibited a licensed dentist from being an employee of an unlicensed business corporation. The court found this to be "reasonable basis" for section 530 relief.

Common law rules

A business that makes a reasonable effort to establish independent contractor treatment for its workers under the common law but falls just short of satisfying the common law standard, may present a valid section 530 safe haven under "other reasonable basis." A reasonable, albeit erroneous, interpretation of the common law rules was found to be sufficient for section 530 relief in *Critical Care Registered Nursing, Inc., supra*.

Prior audit of predecessor

Although a prior audit of the business's predecessor does not satisfy the requirements of the prior audit safe haven, the business may qualify for relief if there has merely been a change in the form of the business. In addition, the successor must be in the same line of business.

PLR/TAM to predecessor

Although a private letter ruling or technical advice memorandum issued to the business's predecessor does not satisfy the requirements of the judicial precedent safe haven, the business may qualify for relief if there has merely been a change in the form of the business.

Good faith

While a number of types of evidence may support a showing of other reasonable basis. In *In re Compass Marine, supra*, the court cited Senate Report No. 1263, 95th Cong. 2d Sess., at 210 (1978), in dicta, as support for the concept that the business has a "reasonable basis" for section 530 relief if it acted in "good faith."

Penalties

Good faith, although not a sufficient basis for section 530 relief, may be a basis for not asserting penalties. *See, Diaz v. United States, supra.*

EFFECT OF SECTION 530 RELIEF ON EMPLOYEE
Status of employee not changed by section 530

As noted previously, section 530 relief does not convert a worker from the status of employee to the status of independent contractor. If it has been determined that worker is an employee, the worker remains an employee for income tax purposes, such as deductions for business expenses and participation in retirement plans.

As previously stated, if the business's liability is terminated by section 530(a)(1), the worker remains liable for employee FICA tax with respect to all wages received. Rev. Proc. 85-18, section 3.08; Treas. Reg. Section 31.3102(c). *See also*, Rev. Rul. 86-111, 1986-2 C.B. 176—The worker remains fully liable for the unwithheld employee FICA tax after the business's liability has been determined under IRC section 3509. The employee's share of FICA tax is reported on Form 4137 by substituting the word "wages" for the word "tips."

SUMMARY
1. Section 530 must be considered as the first step in any worker classification case.

2. Section 530 is a relief provision that has significant impact on the administration of the employment tax laws.

3. Section 530 has been modified, amplified, and defined since 1978 through legislation, IRS revenue rulings, revenue procedures, and court cases. The basic provisions are intact but many interpretation issues remain unresolved.

4. Section 530 provides businesses with relief from federal employment tax obligations if certain requirements are met.

5. The business must meet two consistency requirements before the relief provisions of section 530 apply. For any period after December 31, 1978, the relief applies only if:

• All Forms 1099 required to be filed by the business with respect to the worker(s), for the period, are timely filed and are filed on a basis consistent with the business's treatment of the worker as an independent contractor; and

• The treatment of the worker as an independent contractor is consistent with the treatment by the business (predecessor) of all workers holding substantially similar positions for any period beginning after December 31, 1977.

6. In addition to the consistency requirements, the business must have relied on some reasonable basis, including the safe havens of a prior audit, a judicial precedent, or an industry practice.(Ed. Note: They fail to mention the 20 CLFT).

7. The reasonable basis requirement, including the three safe havens, are to be liberally construed.

8. For examinations beginning before January 1, 1997, a prior audit will provide a safe haven if it is an examination of books and records by the IRS of the same entity, which is still in the same line of business and whose workers are performing substantially the same work. Examinations beginning after December 31, 1996, must have addressed the issue of the status of the class of workers at issue or of a substantially similar class of workers for employment tax purposes.

9. A judicial precedent will provide a safe haven only if the business's case is similar to the precedent. Federal employment tax cases and published rulings qualify. Technical advice memoranda or private letter rulings qualify for the business which requested them. State court decisions and rulings of agencies other than IRS do not qualify.

10. To claim a safe haven under industry practice, the business must show that it is following a long-standing recognized practice of a significant segment of its industry. Industry is the group of businesses that provide the same product or service and compete for the same customers.

11. A business that fails to meet any of the safe havens may be entitled to relief if it can be demonstrated that it relied on some other reasonable basis for not treating the worker as an employee. (Ed Note: such as the 20 CLFT).

Text of Section 530, Including Amendments

I. SECTION 530. CONTROVERSIES INVOLVING WHETHER INDIVIDUALS ARE EMPLOYEES FOR PURPOSES OF THE EMPLOYMENT TAXES.

(a) TERMINATION OF CERTAIN EMPLOYMENT TAX LIABILITY.—

(1) In General.—If—

(A) for purposes of employment taxes, the taxpayer did not treat an individual as an employee for any period, and (B) in the case of periods after December 31, 1978, all Federal tax returns (including information returns) required to be filed by the taxpayer with respect to such individual for such period are filed on a basis consistent with taxpayer's treatment of such individual as not being an employee, then, for purposes of applying such taxes for such period with respect to the taxpayer, the individual shall be deemed not to be an employee unless the taxpayer had no reasonable basis for not treating such individual as an employee.

(2) STATUTORY STANDARDS PROVIDING ONE METHOD OF SATISFYINGTHE REQUIREMENTS OF PARAGRAPH (1).— For purposes of paragraph (1), a taxpayer shall in any case be treated as having a reasonable basis for not treating an individual as an employee for a period if the taxpayer's treatment of such individual for such period was in reasonable reliance on any of the following:

(A) judicial precedent, published rulings, technical advice with respect to the taxpayer, or a letter ruling to the taxpayer;

(B) a past IRS audit of the taxpayer in which there was no assessment attributable to the treatment (for employment tax purposes) of the individuals holding positions substantially similar to the position held by this individual; or

(C) long-standing recognized practice of a significant segment of the industry in which such i individual was engaged.

(3) CONSISTENCY REQUIRED IN THE CASE OF PRIOR TAX TREATMENT.—

Paragraph (1) shall not apply with respect to the treatment of any individual for employment tax purposes for any period ending after December 31, 1978, if the taxpayer (or a predecessor) has treated any individual holding a substantially similar position as an employee for purposes of the employment taxes for any period beginning after December 31, 1977.

(4) REFUND OR CREDIT OF OVERPAYMENT.—If refund or credit of any overpayment of an employment tax resulting from the application of paragraph (1) is not barred on the date of the enactment of the Act by any law or rule of law, the period for filing a claim for refund or credit of such overpayment (to the extent attributable to the application of paragraph (1) shall not expire before the date 1 year after the date of the enactment of this Act.

(b) PROHIBITION AGAINST REGULATIONS AND RULINGS

ON EMPLOYMENT STATUS.—

No regulation or Revenue Ruling shall be published on or after the date of the enactment of this Act and before the effective date of any law hereafter enacted clarifying the employment status of individuals for purposes of the employment tax by the Department of the Treasury (including the IRS) with respect to the employment status of any individual for purposes of the employment taxes.

(c) DEFINITIONS.—For purposes of this section—

(1) EMPLOYMENT TAX.—the term "employment tax" means any tax imposed by subtitle C of the IRC of 1954.

(2) EMPLOYMENT STATUS.—The term "employment status" means the status of an individual, under the usual common law rules applicable in determining the employer-employee relationship, as an employee or as an independent contractor (or other individual who is not an employee).

(d) EXCEPTION.—This section shall not apply in the case of an individual who, pursuant to an arrangement between the taxpayer and another person, provides services for such other person as an engineer, designer, drafter, computer programmer, systems analyst, or other similarly skilled worker engaged in a similar line of work.

RELATIONSHIP OF THE PARTIES

Relationship of business and worker

In this section, we describe other facts that recent court decisions consider relevant in determining worker status. Most of these facts reflect how the worker and the business perceive their relationship to each other. It is much harder to link the facts in this category directly to the right to direct and control how work is to be performed than the categories previously discussed. However, the relationship of the parties is important because it reflects the parties' intent concerning control.

Intent of parties/written contract

Courts often look at the intent of the parties. This is most often embodied in their contractual relationship. Thus, a written agreement describing the worker as an independent contractor is viewed as evidence of the parties' intent that a worker is an independent contractor.

A contractual designation, in and of itself, is not sufficient evidence for determining worker status. The facts and circumstances under which a worker performs services are determinative of the worker's status.

Treas. Reg. section 31.3121(d)-1(a)(3) provides that the designation or description of the parties is immaterial. This means that the substance of the relationship, not the label, governs the worker's status.

The contract may, however, be relevant in ascertaining methods of compensation, expenses that will be incurred, and the rights and obligations of each party with respect to how work is to be performed. In addition, it is difficult, if not impossible, to decide whether a worker is an independent contractor or an employee, the intent of the parties, as reflected in the contractual designation, is an effective way to resolve the issue. The contractual designation of the worker is "very significant in close cases." *See, Illinois Tri-Seal Prods., Inc. v. United States*, 353 F.2d 216, 218 (Ct. Cl. 1965).

Forms W-2 Filing a Form W-2 usually indicates the parties' belief that the worker is an employee. However, workers have succeeded in obtaining independent contractor status even when Forms W-2 were filed. *See, e.g., Butts v. Commissioner,* T.C. Memo 1993-478, *aff 'd per curiam* 49 F.3d 713 (11th Cir. 1995).

Incorporation

Questions sometimes arise concerning whether a worker who creates a corporation through which to perform services can be an employee of a business that engages the corporation. Provided that the corporate\formalities are properly followed and at least one non-tax business purpose exists, the corporate form is generally recognized for both state law and federal law, including federal tax, purposes. Disregarding the corporate entity is generally an extraordinary remedy, applied by most courts only in cases of clear abuse.

Employee benefits

If a worker is excluded from a benefit plan because the worker is not considered an employee by the business, this is relevant (though not conclusive) in determining the worker's status as an independent contractor.

Discharge/termination

The circumstances under which a business or a worker can terminate their relationship have traditionally been considered useful evidence bearing on the status the parties intended the worker to have. Some recent court decisions continue to explore such evidence. However, in order to determine whether the facts before you are relevant to the worker's status, you will need to consider the impact of modern business practices and legal standards governing worker termination.

Discharge/termination—

Traditional analysis

Under a traditional analysis, a business's ability to terminate the work relationship at will, without penalty, provided a highly effective method to control the details of how work was performed and, therefore, tended to indicate employee status. Conversely, in the traditional independent contractor relationship, the business could terminate the relationship only if the worker failed to provide the intended product or service, thus indicating the parties' intent that the business not have the right to control how the work was performed.

Limits on ability to discharge worker

In practice, however, businesses rarely have complete flexibility in discharging an employee. The business may be liable for pay in lieu of notice, severance pay, "golden parachutes," or other forms of compensation when it discharges an employee. In addition, the reasons for which a business can terminate an employee may be limited—whether by law, by contract, or by its own practices. As a result, inability to freely discharge a worker, by itself, no longer constitutes persuasive evidence that the worker is an independent contractor.

Limits on worker's ability to quit

Looking at the issue from the other angle, a worker's ability to terminate work at will was traditionally considered to illustrate that the worker merely provided labor and tended to indicate an employer-employee relationship. In contrast, if the worker terminated work, and the business could refuse payment or sue for nonperformance, this indicated the business's interest in receipt of the product or service for which the parties had contracted and tended to indicate an independent contractor relationship.

Termination of contracts

In practice, however, independent contractors may enter into short-term contracts for which nonperformance remedies are inappropriate or may negotiate limits on their liability for nonperformance. For example, professionals, such as doctors and attorneys, are typically able to terminate their contractual relationship without penalty.

Nonperformance by employee

A business' ability to refuse payment for unsatisfactory work continues to be characteristic of an independent contractor relationship.

FACTS OF LESSER IMPORTANCE
Introduction

This section discusses facts that will typically provide less useful evidence of whether a worker is an independent contractor or an employee. In past decades, these facts were probably more important. However, recent court decisions give them little independent weight. To the extent these facts continue to have relevance, they are generally already reflected in the types of evidence described previously.

Part-time or fulltime work

The fact that a worker performed services on a part-time basis or worked for more than one person or business was once thought to be significant evidence indicating that the worker was an independent contractor. However, in today's economy, whether a worker performs services on a full-time or part-time basis is a neutral fact.

There are several reasons for this change. With cutbacks and downsizing in business and industry, many companies hire workers on a part-time basis. These workers may be either independent contractors or employees. Similarly, working full-time for one business is also consistent with either independent contractor or employee status. An independent contractor may work full-time for one business either because other contracts are lacking, because the contract by its terms requires a full-time, exclusive effort, or because the independent contractor chooses to devote full-time to a particular project.

Finally, many employees "moonlight" by working for a second employer. As a result, whether services are performed for one business is no longer useful evidence.

Place of work.

Whether work is performed on the business's premises or at a location selected by the business often has no bearing on worker status. Even when it i s relevant evidence, it will be relevant because it illustrates the business's right to direct and control how the work is performed and will have been considered in connection with instructions.

One location

In many cases, services can be provided at only one location. For example, repairing a leaky pipe requires a plumber to visit the premises where the pipe is located. Similarly, a camera operator must shoot a commercial at the same location as the director and actors. These requirements are inherent in the result to be achieved and are not evidence of the right to direct and control how the work is performed.

Different locations

In other cases, work can be performed at many different locations. Modern technology has developed tools that greatly expand the scope of the workplace, such as cellular phones, modems, and computer networks. Allowing work off site can be attractive to businesses due to lowering costs, improving morale, and helping to retain valued workers. In today's world, off-site work is consistent with either an independent contractor or employer-employee relationship.

The place where work is performed is most likely to be relevant evidencein cases in which the worker has an office or other business location. However, you will have already considered this evidence in evaluating significant investment, unreimbursed expenses, and opportunity for profit or loss.

Hours of work

You can easily apply the same reasoning that we used in connection with place of work to understand why hours of work is also a fact that, if relevant, has already been considered in connection with instructions. Some work must, by its nature, be performed at a specific time.

Again, our camera operator must be ready to provide photography services when the director and actors are on hand. This relates to the result to be achieved, not how the work is performed. Modern communications have increased the ease of performing work outside normal business hours, while flexibility in setting hours may improve morale and retain valued workers. In today's world, flexible hours are consistent with either independent contractor or employee status.

WEIGHING THE EVIDENCE

Control and autonomy both present

When you have explored the relevant evidence, you will probably find some facts that support independent contractor status and other facts that support employee status. This is because independent contractors are rarely totally unconstrained in the performance of their contracts, while employees almost always have some degree of autonomy. Which predominates?

You will, therefore, need to weigh the evidence before you in order to determine whether, looking at the relationship as a whole, evidence of control or autonomy predominates. You may, for example, find that the business requires the worker to be on site during normal business hours, but has no right to control other aspects of how the work is to be performed; that the worker has a substantial investment and unreimbursed expenses combined with a flat fee payment; and that contractual provisions clearly show the parties' intent that the worker be an independent contractor. In this case, you would logically conclude that the worker was an independent contractor despite the instructions about the hours and place of work.

SUMMARY

Review of lesson The following summarizes what we have covered in this lesson:

1. In determining a worker's status, you should gain an understanding of the way a business operates and the relationship between the business and the worker.

2. Areas to consider while developing your case are:

• What the business does and how the job gets done.

• The relationship between the business and its clients or customers.

• Facts that indicate whether the business has the right to control how work is done.

3. Evidence that may be the most persuasive can be identified within three specific categories.

• Behavioral control.

• Financial control.

• Relationship of the parties.

4. Behavioral control focuses on whether there is a right to direct or control how the work is done. The presence or absence of instructions and training on how work is to be done are especially relevant.

5. Financial control focuses on whether there is a right to direct or control how the business aspects of the worker's activities are conducted. Significant investment, unreimbursed expenses, services available to the relevant market, method of payment, and opportunity for profit or loss are facts relevant to financial control.

6. Relationship of the parties focuses on how the parties perceive their relationship. Intent of parties/written contract, employee benefits, discharge/termination, permanency, and regular business activity are relevant to how the parties perceive their relationship.

7. Relevant evidence in all three categories must be weighed to determine the worker's status.

SELECTED CASE

Illinois Tri-Seal Products v. United States, 353 F.2d 216 (Ct. Cl. 1965) An excellent history of the Congressional repudiation of the *Silk/ Bartels* "economic reality" approach can be found in this case holding window installers to be independent contractors. The case also illustrates the intrinsically factual nature of independent contractor/employee determinations. It also contains helpful discussions of the distinction between instructions and suggestions and of the significance of the parties' view of their relationship in close cases.

STATUTORY EMPLOYEES, STATUTORY NON-EMPLOYEES, AND OTHER CLASSES OF WORKERS

INTRODUCTION

In the previous lesson, you studied what constitutes a common law employee where the business is liable for FICA, FUTA, and federal income tax withholding. In this lesson, you will study corporate officers and certain workers that are defined by statute as employees, commonly referred to as "statutory employees." You will also study workers in three occupations where, by statute, the worker performing the services is specifically not treated as an employee (commonly referred to as "statutory non-employee").

In cases of a statutory non-employee, the business for which the services are performed is not treated as an employer, and, therefore, is not liable for any of these taxes. IRC section 3121(d) contains four categories of employees for FICA tax purposes:

- common law employees
- corporate officers
- statutory employees
- employees covered by an agreement under section 218 of the Social Security Act IRC section 3508 contains tests for the treatment of real estate agents and direct sellers as statutory non-employees. IRC section 3506 provides the requirements for treating companion sitters as statutory non-employees.

1. Determine whether a corporate officer is an employee for purposes of FICA and FUTA taxes and federal income tax

withholding.

2. Identify statutory employees for purposes of FICA and FUTA taxes.

3. Identify statutory non-employees.

CORPORATE OFFICERS

Exception

Officers are specifically included within the definition of employee for purposes of FICA, FUTA, and federal income tax withholding. See IRC sections 3121(d)(1), 3306(i), and 3401(c). The common law standard is not applicable. The regulations provide that generally an officer of a corporation is an employee of the corporation.

However, an officer is not considered to be an employee of the corporation if two requirements are met: (1) the officer does not perform any services or performs only minor services; and (2) the officer is not entitled to receive, directly or indirectly, any remuneration. Treas. Reg. section 31.3121(d)-1(b). The officer must meet both requirements to be excepted from employee status. In determining whether services performed by a corporate officer are considered minor or nominal, examine the character of the service, the frequency and duration of performance, and the actual or potential importance or necessity of the services in relation to the conduct of the corporation's business.

A director of a corporation, acting in the capacity of a director, is not an employee of the corporation for those services, even if that worker also serves as an employee or officer of the corporation for other services. Therefore, part of the compensation paid this worker can be for services rendered as an independent contractor (director) and part of the payments can be for services rendered as an employee. Rev. Rul. 58-505.

EXAMPLE

Various officers of five related operating corporations performed only minor ministerial functions entailing a few hours work a year for the corporations.

The officers also received no remuneration for the services they performed for these five corporations. Because the officers satisfied both requirements for the exception from employee status (*i.e.*, they performed only minor services for the corporations and received no remuneration), the officers were not employees of the operating corporations. Rev. Rul. 74-390, 1974-2 C.B. 331

Payments to officers

You should closely examine all payments to the officer, such as amounts labeled as draws, loans, dividends, or other distributions, to determine whether the payments are in fact wages for FICA, FUTA, and federal income tax withholding purposes.

STATUTORY EMPLOYEES

Statutory employee

If a worker is not an employee under the usual common law rules or a corporate officer, the worker and the business may nevertheless still be subject to employment taxes. IRC section 3121(d)(3) lists workers in four occupational groups who, under certain circumstances, are considered employees for FICA tax, and, in some instances, FUTA tax, but not for federal income tax withholding. These groups include:

- agent-drivers or commission-drivers
- full-time life insurance sales persons
- home workers
- traveling or city sales persons

These workers are referred to as "statutory employees." Workers in these four occupational groups are employees for FICA tax purposes. By definition, a worker cannot be a statutory employee under IRC section 3121(d)(3) if that worker is a common law employee. *See Lickiss v. Commissioner*, T.C. Memo 1994-103.

General requirements

In order for IRC section 3121(d)(3) to apply when a worker performs services for remuneration for a business, there are three general requirements.

They are:

1. The contract of service contemplates that the worker will personally perform substantially all the work.

2. The worker has no substantial investment in facilities other than transportation facilities used in performing the work.

3. There is a continuing work relationship with the business for which the services are performed.

Work performed personally

The term "contract of service" means an arrangement oral or written, under which the particular services are performed. The term "personally perform" means it is contemplated that the worker will do substantially all the work personally. Therefore, if the arrangement contemplates that the worker would be free to delegate as much of the work as he or she desires, then the worker could not be a statutory employee under this section.

No substantial investment

The term "substantial investment" is not defined in the regulations. All of the facts for each case must be considered to determine whether the facilities furnished by the worker are substantial. Several factors listed below should be considered:

1. What is the value of the worker's investment compared to the total investment?

2. Are the facilities furnished essential to perform the work or for the personal convenience of the worker?

3. Are the facilities being purchased or leased at fair market or fair rental value?

4. Are the facilities furnished by the worker considerably more extensive than those usually furnished by workers performing comparable services?

Continuing relationship

Work is considered to be of a continuing nature if it is regular or frequently recurring. Regular part-time and regular seasonal work are considered continuing. A single job transaction is not generally a continuing relationship.

CATEGORIES OF STATUTORY EMPLOYEES LIABLE FOR FICA BUT NOT WITHOLDING (PAYROLL TAX).

Agent drivers or commission drivers

The statute limits agent drivers or commission drivers to workers who distribute meat or meat products, vegetables or vegetable products, fruit or fruit products, bakery products, beverages (other than milk), or laundry or dry-cleaning services for a business. The distribution of other services or products will not disqualify the worker from this category of statutory employee if handling the additional products or services is incidental to handling the specified items.

The agent or commission drivers may sell at retail or wholesale. They may operate from their own trucks or from trucks belonging to the business for which they work. The drivers may serve customers designated by the business as well as those they solicit.

Their compensation may be based on commission, or the difference between the price charged to the customer and the price paid by the driver to the business for the product or service.

Full-time life insurance salespersons

This group includes salespersons whose full-time occupation is soliciting life insurance or annuity contracts or both, primarily for one life insurance business. Generally, the contract of employment reflects the intent of the worker and the business in determining whether the worker is a full-time or part-time salesperson. The actual time devoted to the work is not determinative. A worker may work regularly only a few hours each day and still qualify as a full-time life insurance salesperson.

The salesperson's efforts must be devoted primarily to soliciting life insurance or annuity contracts. Occasional or incidental sales of other types of insurance for the business, or the occasional placing of surplus-line insurance, will not affect this requirement. However, the salesperson who devotes substantial efforts to selling applications for insurance contracts other than life insurance and annuity contracts (for example, health and accident, fire, automobile, etc.) does not meet the requirement.

Home workers

The term "home worker" can encompass workers who perform a wide range of duties. Traditionally, this group would have included, but was not limited to, workers who would make such things as clothing, bedding, needlecraft products, or similar products

Specific requirements for home workers

To qualify as a statutory employee, the worke r must meet, in addition to the three general requirements previously listed, the following requirements:

1. The work must be done in accordance with the specifications given by the business (generally, simple and consisting of such things as patterns or samples).

2. The material or goods on which the work is done must be furnished by the business.

3. The finished product must be returned to the business or to another designation. It is immaterial whether the business picks up the work, or the worker delivers it.

Traveling or city salesperson

This category includes workers who operate away from the

business's premises. Their full-time business activity is selling merchandise for a business. The test of full-time relates to an exclusive or principal business activity for a single business and not to the time spent on a job. Sideline sales activities for some other business, however, do not exclude these workers from coverage.

Specific requirements for traveling or city salespersons

In order for traveling or city salespersons to fall within the statutory test, they must meet, in addition to the three general requirements previously listed, the following requirements:

1. Their entire or principal business activity must be devoted to soliciting and transmitting orders for merchandise of a single business.

2. The orders must be obtained from wholesalers, retailers, contractors, or operators of hotels, restaurants, or other similar establishments.

3. The merchandise sold must be bought for resale or must be supplied for use in the purchaser's business operations.

Principal business activity defined for traveling or city salespersons. Generally, the test is met if 80 percent of the activity is for one business.

Types of purchasers for traveling or city salespersons

Workers must sell principally to the classes of purchasers described in IRC section 3121(d)(3)(D) to be considered statutory employees and liable for FICA. They may also sell incidentally to others.

CLASS OF PURCHASER DESCRIPTION
Wholesaler

A wholesaler buys merchandise in large quantities and usually sells in small quantities to jobbers or to retail dealers but not to the ultimate consumer. The wholesaler does not process the merchandise in any way to cause it to lose or change its identity.

Retailer

A retailer buys merchandise in small quantities and then sells it in smaller quantities usually to the ultimate consumer. Retail establishments may perform service functions or processing or manufacturing operations with respect to the items they sell without losing their character as retail establishments. For example, a store which sells drapery and slip cover material, and also makes draperies and slip covers for the consumer, is a retail establishment and not a manufacturer. A neighborhood bakery is essentially a retail store, even though it changes the form of the raw material to the final prepared material. Contractors include such service organizations as contractors for window washing, wall cleaning, construction, and other services.

Hotels, Restaurants, or Other Similar Establishments

The phrase "other similar establishments" refers solely to establishments similar to hotels and restaurants whose primary function is the furnishing of food or lodging.

Classes of purchasers not included for traveling or city salespersons

Manufacturers, schools, hospitals, churches, municipalities, and state and federal governments are not within the included classes of purchasers. A manufacturer produces articles for use from raw or prepared materials by giving them new forms, qualities, and properties, or combinations of these items. Sales made to a unit of an organization not within the included classes of purchasers may meet the requirements regarding "classes of purchasers" provided the unit carries on a separate and clearly identifiable business with a type of purchaser described in IRC section 121(d)(3)(D).

For example, sales made to an unincorporated university bookstore, owned and operated by the university, are sales made to a purchaser included in the statutory definition of "traveling or city salesperson."

Resale or use for traveling or city salespersons

Merchandise must be for resale or for use in the business operation of the purchaser. The phrase "merchandise for resale" includes only tangibles which do not lose their identity as they pass through the hands of the purchaser. The phrase "supplies for use in the business operation" means principally supplies used in conducting the purchaser's business. This includes all tangible merchandise not considered "merchandise for resale." Services, such as radio time and advertising space, are intangible and outside the definition. However, items such as advertising novelties and calendars constitute supplies within the definition. Service may be part of the sale for traveling or city salespersons

If workers perform substantial work in servicing the articles they sell, they may still meet the requirements of IRC section 3121(d)(3)(D). For example, a worker who spends a day selling a machine and a day supervising its installation and training the purchaser's personnel in its use may still have performed services as a full-time salesperson. Furnishing such services may be a necessary part of the inducement for the buyer to purchase. The question, therefore, is whether the total activity is essentially a selling activity. If so, the services related to the sale, even though substantial, are an integral part of the sale.

Statutory employees' expenses

Statutory employees under IRC section 3121(d)(3) are not employees for the purpose of deducting trade or business expenses. Therefore, they may deduct their expenses on Schedule C rather than as miscellaneous itemized deductions. Rev. Rul. 90-93, 1990-2 C.B. 33.

Statutory employees receive a Form W-2. A check is made in Box 15 to indicate that the worker is a statutory employee. Federal income tax withholding does not apply to statutory employees. If statutory employees also have earnings from self-employment, they may not use expenses from services as a statutory employee to reduce net earnings from self-employment for purposes of SECA, IRC section 1402(a). This is because services as a statutory employee do not constitute the carrying on of a trade or business for purposes of SECA. Statutory employees are required to file a Schedule C for services performed as a statutory employee separate from a Schedule C that reports net earnings from self-employment.

Statutory employee treatment

Recently, workers who were otherwise common law employees have claimed to be statutory employees to be eligible for the treatment of Rev. Rul. 90-93. The issue in Rev. Rul. 90-93 was whether a full-time life insurance salesperson who was treated as an employee for FICA purposes under IRC section 3121(d)(3) was also an employee for purposes of IRC sections 62 (relating to above the line deductions) and 67 (relating to two percent floor on miscellaneous itemized deductions). The holding was that a full-time life insurance sales person described in IRC section 3121(d)(3) is not an employee for purposes of sections 62 and 67. Rev. Rul. 90-93 also applied to all other statutory employees described in IRC section 3121(d)(3) in connection with expenses they incur in the conduct of their trades or businesses.

If a worker's return appears to take inconsistent positions, further evaluation is appropriate. For example, if a worker's return includes a W-2 indicating employee status yet claims deductions related to this income on Schedule C, you should ask the worker for an explanation of the potentially inconsistent positions. If the worker is not a statutory employee, the appropriate adjustment should be made.

Remember, the worker can be a statutory employee only if the worker is an independent contractor under the common law standard.

Statutory employee benefit plans

Except for full-time life insurance salespersons, statutory employees under IRC section 3121(d)(3) remain independent contractors for employee benefit purposes. Thus, they are not eligible to participate in the employee benefit plans sponsored by the business for employees and cannot enjoy the exclusions from income for amounts paid under accident and health insurance arrangements under IRC sections 104, 105, and 106 to the extent that those income tax exclusions apply only to employees. However, statutory employees can establish and maintain their own self-employed retirement plans.

Full-time life insurance salespersons are an exception. They are treated as employees not only for FICA tax purposes, but also for certain employee benefit programs maintained by the business. IRC section 7701(a)(20). Thus, they may participate as employees under the business's group term life insurance program under IRC section 79, apply the exclusions available to employees participating in the business's accident and health plans under IRC sections 104, 105, and 106, apply the exclusion from income under IRC section 101(b) for employer provided death benefits, and participate as an employee in the business's qualified deferred compensation or retirement plans under IRC section 401(a) and the business's cafeteria plan under IRC section 125.

On the other hand, a full-time life insurance salesperson may not base contributions to a self-employed retirement plan (commonly called a Keogh plan) on the compensation received from the insurance business.

STATE AND LOCAL GOVERNMENT EMPLOYEES
218 Agreement

IRC section 3121(d)(4) provides that workers for state and local governments are employees for FICA purposes if the governmental unit has entered into an agreement with the Social Security Administration to provide FICA coverage pursuant to Section 218 of the Social Security Act.

These agreements may be broad or may deal with very specific worker groups. Since April 20, 1983, coverage under a 218 agreement cannot be terminated. They can be employees for FICA purposes under common law. As a result of legislative changes since 1986, workers for state and local governments can also be employees for FICA purposes if they are employees under the common law rules, even though the worker's services are not covered under a Section 218 Agreement.

In analyzing how workers who are not covered under a Section 218 Agreement are treated, it is helpful to keep in mind that FICA taxes consist of two components, Old Age, Survivors, and Disability Insurance (OASDI) and Hospital Insurance (Medicare). For services performed after July 1, 1991, both the OASDI and the Medicare components of FICA apply to state and local government common law employees, unless the employee is covered by a public retirement system. As most of these governments have broad coverage in their public retirement systems, relatively few state and local government employees are covered by this rule.

The Medicare portion of FICA taxes applies to wages of state and local government common law employees hired after March 31, 1986, unless the employee meets the continuing employment exception of IRC section 3121(u)(2)(C).

STATUTORY NON-EMPLOYEES
Introduction

Workers in three occupations will not be treated as employees for FICA, FUTA, or federal income tax withholding purposes provided they meet certain qualifications. These workers are referred to as "statutory non-employees." IRC section 3508 provides that, for all IRC purposes, qualified real estate agents and direct sellers are statutory nonemployees. IRC section 3506 provides that, for purposes of subtitle C of the IRC relating to employment tax, (FICA and FUTA taxes, and federal income tax withholding), qualifying companion sitters are statutory nonemployees.

Qualified real estate agents

IRC section 3508 provides that a worker is a qualified real estate agent if the following requirements are met:

1. The worker is a licensed real estate agent.

2. Substantially all of such worker's remuneration for services is directly related to sales or other output rather than to the number of hours worked.

3. A written contract exists between the worker and the business for which services are being performed that provides that the worker will not be treated as an employee for federal tax purposes.

Direct sellers IRC section 3508 provides that a worker is a direct seller if the following qualifications are met:

1. The worker is engaged in the sale of consumer products in the home or in other than a permanent retail establishment.

2. Substantially all of such worker's remuneration for services is directly related to sales or other output rather than the number of hours worked.

3. A written contract exists between the worker and the business for which the services are being performed that provides that the worker will not be treated as an employee for federal tax purposes. The proposed regulations drafted for IRC section 3508 include detailed explanations of the terms used to define "direct seller."

Since their publication in 1986, the regulations have come under increasing criticism. One area that has been attacked concerns the definition of "consumer products." Proposed Treas. Reg. section 31.3508-1(g)(3) defines "consumer products" as tangible personal property used for personal, family, or household purposes, including property intended to be attached to, or installed in any real property.

Litigation of definition of consumer products

The definition of "consumer products" was litigated in *Cleveland Institute of Electronics, Inc.*, 787 F.Supp. 741 (DC. ND. Ohio 1992). In this case, the products being sold were home study educational courses. The Service deemed these courses to be intangible in nature and consequently held that they did not meet the proposed regulations' definition of "consumer products." The District Court's consideration of this matter resulted in the conclusion that the proposed regulations' definition of "consumer products" was unnecessarily restrictive. In deciding that the workers selling these courses were independent contractors, the court expanded the definition to include not only tangible consumer goods, but also intangible consumer services such as the courses at issue.

Consumer products definition expanded

This expanded definition of "consumer products" was subsequently cited in *The R Corporation*, 94-2 USTC par. 50,380 (DC. M.D. FL (1994)) where sellers of TV cable services were involved. In ruling that the sellers of cable service were direct sellers under IRC section 3508, the court concluded that the cable service being sold was an intangible consumer product.

Based upon the litigation cited above, and pending finalization of the regulations and further consideration of this issue in that context, cases should not be developed based on a distinction between tangible and intangible products; *i.e.*, both types of products will qualify. In your consideration of direct seller cases, care should be taken, therefore, to ensure that your research is current.

Newspaper carriers and distributors

The Small Business Job Protection Act added qualifying newspaper distributors and carriers as direct sellers. Under the amendment, a person engaged in the trade or business of the delivery or distribution of newspapers or shopping news qualifies as a direct seller provided all remuneration is directly related to sales or output, rather than hours worked. Also, the services must be performed pursuant to a written contract that provides the person will not be treated as an employee for Federal tax purposes. The provision is effective with respect to services performed after December 31, 1995.

Companion sitters

IRC section 3506 provides that a companion sitter will not be an employee of a companion sitting placement service if the companion sitting placement service neither pays nor receives the salary or wages of the sitter. The placement service may be compensated on a fee basis by either the sitter or the person or business for which the sitting is performed.

The companion sitter is deemed to be self-employed unless considered to be a statutory or common law employee of t he person or business for which the services are performed. Treas. Reg. section 31-3506-1(c) and (d).

SUMMARY

The following summarizes what we have covered in this lesson:

1. Officers of corporations are employees for purposes of FICA,FUTA, and federal income tax withholding unless the services rendered are minor or nominal, and they neither receive nor are entitled to receive any compensation.

2. Certain classes of workers who do not meet the common law rules of determining employer-employee relationships are still employees for FICA tax purposes. These statutory employees are:

• Agent or commission drivers.

- Full-time life insurance salespersons.
- Home workers.
- Traveling or city salespersons.

3. Before a worker in one of these four categories is considered a statutory employee, three general requirements must be met:

Contract of service states that the work will be performed personally.

Worker has no substantial investment in facilities.

Continuing relationship exists between the worker and the business.

4. In determining whether a worker is an employee, first apply the common law rules. If the facts do not support a position that a worker is a common law employee, then apply a test to determine if the worker is a statutory employee. If you determine that a statutory employee situation exists, the business is liable for FICA tax.

The business may also be liable for FUTA tax. Remember, the determination that a worker is a statutory employee is made for employment tax purposes. Thus, the worker is not subject to federal income tax withholding and is not eligible for voluntary federal income tax withholding because the common law status is that of independent contractor, not employee.

5. Qualified real estate agents, direct sellers, including newspaper carriers and distributors, and companion sitters are statutory nonemployees and are not treated as employees for purposes of FICA, FUTA, and federal income tax withholding, if they meet certain IRC requirements.

AUTHOR'S BIOGRAPHY

SHELDON R. WAXMAN
BOX 309
SOUTH HAVEN MI 49090-0309
(269) 207-6219
E-MAIL: sheldonw72@gmail.com
WEBSITE www.thelawyer.info, www.independentcontractor.com

PERSONAL:
• Born: Chicago, Illinois, April 22, 1941.
• Married: Katherine.
• Children: Josiah and Zoe .

EDUCATION:
• University of Illinois, B.A. (1963). I Majored in Political Science and had a split Minor in Philosophy and Comparative Religion—these are subjects which I followed through my life's path and, also, an ever-abiding interest in American History and Economics;
• DePaul University College of Law, J.D. (1965).
SCHOLASTIC HONOR: Member, Blue Key National Honor Fraternity.
ACADEMIC HONOR: Case note Editor, DePaul Law Review.
• Post-College: Adv. Diplomate, Court Practice Inst. of Chicago.
BAR ADMISSIONS
• United States Supreme Court (1976),
• Courts of Appeal for 6th, 7th, 9th and 10th Circuits,
• Trial Bar-U.S. District Court for the Northern District of Illinois,
• U.S. District Court for the Western District of Michigan,
• Supreme Courts of Illinois (1965) and Michigan (1985).

SPECIALIZED BACKGROUND
• Former Member, Employment Tax Committee of the Tax Section of the American Bar Association,
• Former Member, Tax Section, Michigan Bar Association,
• Former Member, Tax Freedom Institute,

• Former Trustee in Bankruptcy,

• Former Federal Defender Panel Member (Chicago, Grand Rapids),

• Former Receptionist, Executive Club of Chicago Reception Committee,

• Appearance before Senator Montoya's Special Hearings Committee on IRS Abuses, 1978,

• House Ways and Means Committee Appearance on New Tax Proposals—urging adoption of a Flat Rate Income Tax or a National Sales Tax, 1978,

• Staff Meeting, Joint Committee on Taxation-regarding Section 530 of the Revenue Act of 1978,

• Q Clearance-Atomic Energy Commission,

• Confidential Clearance Federal Bureau of Investigation,

• Marquis', Who's Who in the World and Who's Who in American Law ,

• Former-Michigan Assigned Appellate Counsel,

• Van Buren County, Michigan Court Appointed Defense Lawyer for i ndigents,

• Civil Mediator (Van Buren, Kalamazoo & Berrien Counties),

• Former Member, Mediation Committee Kalamazoo Bar Association,

• Former Member, DePaul University Alumni Association High School Representative,

• Speaker, American College of Trial Lawyers (1997),

• Seminar Speaker, Independent Contractor Issue and other legal issues and libertarian philosophy.

TEACHING EXPERIENCE
• Taught course in Torts for the American Institute for Paralegal Studies, Inc. (1989),

• Lectured on Poverty Law at DePaul University Law School.

• Adjunct Instructor, Political Science, Grand Rapids Community College (1999).

PAST EMPLOYMENT
• Assistant U.S. Attorney (Chicago-1971-1974),

• Staff Attorney, Argonne IL., National Laboratory (1968-1971),

• Supervisory Attorney, Montrose Legal Aid Bureau (Chicago-1966-1968),

• Attorney, Edwin F. Mandel Legal Aid Clinic of the University of Chicago Law School (1965-1966).

ACADEMIC PUBLICATIONS

• Casenote, "Real Property—Business Compulsion: Exaction of Overpayment", 12 DePaul Law Review 351 (1963),

• Comment, "Attorney-Client Privilege: Does it Apply to Corporations?" 12 DePaul Law Review 263 (1963),

• Co-authored Articles, Trial Magazine, "The Federal Preliminary Examination," ATLA (Spring, 1979) also appearing in Barrister and Law Notes of the American Bar Association,

• Lincoln Award, Illinois State Bar Association writing contest, "Conflict in Illinois Courts on Choice of Law Theory—Is It Lex Loci Delictus or Substantial Interest" (Nov., 1977 Illinois Bar Journal),

• Articles in Decalogue Journal, Vol. 24, No. 2 at p. 8 (Fall/Winter, 1976-77) "State Liability for Commission of Federal Civil Rights Violations by Its Employees" and Vol. 26 at p. 16 (Winter/Spring 1979-80) "Tax Complexity vs. Simplicity."

OTHER PUBLICATIONS

• Book Review, "Julian Jaynes, The Origin of Consciousness in The Breakdown of the Bicameral Mind," Vol. 434 Annals, American Academy of Political and Social Science, p. 239 (Nov. 1977),

• "In the Teeth of the Wind-A Study of Power and How to Fight It" (2002),

• Editor, "New Z Letter." (1980-81).

SAMPLE OF SOME NOTEWORTHY REPORTED CASES

• Marchlik v.Coronet Insurance Co., 239 N.E. 2d 299 (IL Supreme Court, 1968),

• Vitale v. INS., 463 F.2d 579 (7th Cir. 1972),

• Ramirez v. Weinberger, 365 F. Supp. 105 (3 Judge, N.D. IL. 1973) aff'd Per Curiam, 415 U.S., 970,

• U.S. v. Lomar Discount, Ltd., 61 F.R.D. 420 aff'd., 498 F.2d 1404 (7th Cir. 1974),

• U.S.v. Dema, 544 F. 2d 1373 (1976),

• U.S.ex. rel. Stewart v. Scott, 501 F.Supp. 53 (N.D. IL., 1980),

• Powe v. City of Chicago and County of Cook, 664 F.2d 639 (7th Cir. 1982),

• Dema v. Feddor, 470 F.Supp. 152 (N.D. IL. 1979) aff'd., 661 F.2d 936,

• Tabcor Sales Clearing, Inc. v. Department of Treasury, 471 F.Supp. 436 (N.D. IL., 1979),

• Illinois v. Gorman, 421 N.E. 2d 228 and 444 N.E. 2d 776 (1st District, IL. App. Ct., 1981 and 1982),

• Lipsey v. Chicago Cook County Criminal Justice Commission, 629 F.Supp. 955 (N.D. IL., 1986),

• Lipsey v. Illinois Human Rights Commission, 510 N.E. 2d 1226 (1st District, IL., App.Ct., 1987),

• Vogel v. Social Security Administration, 735 F.Supp.1353 (N.D. IL. 1989),

• Chief Trial Attorney, Michigan v. Fisher, 483 N.W. 2d 452 (Mi. App. Ct., 1992),

• Norwood Industries, Inc. v. Grand Blanc Printing, Inc., 556 N. W. 2d 897 (Mi. App. Ct., 1996).

HISTORY OF PRACTICE

• Since 1975, I have been engaged in the sole practice of law, along with associates throughout the United States, who are consulted on a need basis. We have a boutique practice. Our emphasis has been on Major and Complex civil and criminal, state and federal litigation, Federal Tax and Constitutional litigation. (Revised, 2011)

Other Books by Author:
"Rebel In the Courtroom"
"The Case Adventures of Sam Cohen, J.D."

INDEPENDENT CONTRACTOR

In the summer of 2011, Shelly donated 30 boxes of his papers and 200 books to the Wilcox Collection of Contemporary Political Movements of the Spencer Research Library at the Universtiy of Kansas. The catalogued collection may be viewed by appointment.

Shelly can be contacted at sheldonw72@gmail.com.
Visit, also, his websites
www.thelawyer.info
www.independentcontractor.info
www.postpubco.com/shelly.htm

Also by Sheldon "Shelly" Waxman, J.D:

With co-author the iconoclast, provocative novelist, and futurist **James Nathan Post**, Shelly writes these tales of the adventurous cases of a hard-working, street-wise, but still idealistic maverick libertarian Chicago lawyer named Sam Cohen. Published under the imprint NewPulp, these are tough, hardcore and hardcorps, taboo-busting stuff that will make you think. You'll recognize some of the cases, as they are based on chapters of this book.

These are available in paperback and on Kindle. The best deal is the new trilogy edition:

THE CASE ADVENTURES OF SAM COHEN, J.D.

which includes:

THE BLACK MESSIAH MURDERS
For the midnight raid in which Black Panther Fred Hampton was killed. Was William O'Neal the black Donnie Brasco of the FBI, or the Judas of the Black Panther Party?

PIRANHAS ON THE LOOSE
Sedlack was an Army assassin who loved his job. Why was he in Chicago's quasi-legal anti-protester Legion Of Justice?

THE JOSEPHUS ENIGMA
Why was someone staging sacrificial murders in Chicago? How were they using the courts and the media to cover them up, and why were they targeting Sam Cohen's client Andy McAndrew?

REBEL IN THE COURTROOM

A fast-reading factual account of some of Shelly's most interesting and challenging cases.

www.ingramcontent.com/pod-product-compliance
Lightning Source LLC
Chambersburg PA
CBHW071242170526
45165CB00003B/1209